A REGIONAL HISTORY OF
THE RAILWAYS OF GREAT BRITAIN

Volume XIII
THAMES AND SEVERN

A REGIONAL HISTORY OF
THE RAILWAYS OF GREAT BRITAIN

Volume 13

THAMES AND SEVERN

Rex Christiansen

WITH 32 PLATES
11 MAPS
AND FOLDING MAP

DAVID & CHARLES
NEWTON ABBOT   LONDON   NORTH POMFRET (VT)

British Library Cataloguing in Publication Data
Christiansen, Rex
    Thames and Severn. – (A Regional History
    of the Railways of Great Britain; v.13)
    1. Railways – England – Thames Valley –
    History
2. Railways – England – Severn Valley –
    History
    I. Title   II. Series
    385'.09424   HE3019.T/
    ISBN 0 7153 8004 4

Typeset by ABM Typographics Limited, Hull
and printed in Great Britain
by Butler & Tanner, London & Frome
for David & Charles (Publishers) Limited
Brunel House   Newton Abbot   Devon

Published in the United States of America
by David & Charles Inc
North Pomfret   Vermont 05053   USA

Contents

Note on folding map, inside back cover:
The folding map has been drawn to show a wider area
   to give a comprehensive idea of the main lines that
   ran *through* the Thames & Severn region, rather than
   cutting them at the boundaries of this volume. Thus
   lines like those from Paddington to the West and
   Bristol to Birmingham include the detail of connecting lines at the extremeties which are described in
   other volumes in this series.

CHAPTER I

A Region of character and variety

Listening to the labouring beat of our double-headed holiday express climbing out of the Severn Tunnel towards Bristol bound for the West of England was one of the incomparable joys of childhood. A long wait to get through Temple Meads, which inevitably began a few minutes later, was one of the frustrations of family holidays, vividly remembered from the 1930s.

Our express from Liverpool (where father gave money to young boys running bare-footed on waste land we crossed to get to Lime Street station), had run via Crewe, Shrewsbury and the Welsh border, where the hills towering above the line at Church Stretton and Abergavenny introduced me to the glories of a countryside I have loved ever since.

Few stations have been more aptly named than Temple Meads, for apart from its suggestive design it was surely the temple of the Great Western Railway, whose title deserves savouring in full to give full emphasis to that great railway. We may dismiss any claim by Paddington to have been the GWR's 'temple' since it was never the grand junction that Bristol was and still is to some degree.

'The fame of Bristol merchant ships and Bristol merchants extends back to unrecorded time,' as Charles Wells pointed out in *A Short History of the Port of Bristol* of 1908 – that was anything but short since it ran to more than 400 pages. He stated that by 1700 Bristol was second to London as England's largest city and port, its wealth based largely on trade with America and the West Indies. Much of its importance was due to its sheltered position on Severnside, for the river, with canal, gave access to Gloucester and Birmingham. And before 1800, Thames and Severn were linked by canal. First major expansion at Bristol took place between 1803–9 when three miles of tidal river were converted into the Floating Harbour or City Docks.

Restored Regency terraces and squares are pleasing reminders of the City's wealth before railways arrived, the first major step being the opening of the London main line in 1841. By Edwardian times Bristol's population was well past 300,000 — halfway towards its present total. It is a region rather than just a city, of constant change, its traditional industries including tobacco, wine and chocolate, complemented by aircraft manufacture (of which the Concorde supersonic airliner is perhaps the best-known product), and chemicals, based largely at Avonmouth.

In the City centre, office accommodation has quadrupled in recent years, partly because banking and insurance companies have found residential, recreational and educational facilities good enough to tempt staff to move out of London.

Transport facilities could be added to that list, not least because Bristol benefits from two High-Speed Train services, serving Temple Meads and Parkway, the latter imaginatively sited since 1 May 1972 (four years ahead of the HST services) on the South Wales main line at an exposed and windy, but commercially viable spot close to two long-distance motorways (M4 and M5) and a local one (M32) serving Central Bristol. Parkway's planners anticipated population growth north of Bristol and in the Severnside/Gloucestershire areas rising from 178,000 in 1968 to a present-day total of just over a quarter of a million. By 1974, when 1,000 London-bound commuters a day were using Parkway, it was transformed from a virtually basic station to something a little more comfortable by weather protection for its two platforms and overbridge.

Parkway has proved to be a station popular with old people, travellers not used to long-distance journeys, and those who often get baffled and flustered and do not relish changes at large stations like Temple Meads. The elderly travelling between the North and the Midlands to South Wales like Parkway because they do not have to change platforms when changing trains.

Bristol is perhaps one of the largest centres of population of its kind—widely separated from others—with such light commuter traffic. In 1979 a survey showed that the total number using the Severn Beach branch and the Bath–Bristol–Weston-super-Mare corridor daily was no more than 1,200. Despite that small total city-based architects and engineers presented a plan for an Avon Metro with a central interchange at Lawrence Hill for routes, mostly underground in the city centre, radiating to Brislington and South Bristol, Weston, Portishead, Severn Beach, Yate, and Bath via two routes: direct and via Warmley.

Initial estimates for the Avon Metro put the cost at £200 million, but Bristol's Euro-MP, Richard Cottrell, claimed in November 1979 that the City would enter the 21st Century with a transport system considerably worse than that with which it left the 19th Century.

But it looks as though the County's road rather than rail system is set for possible expansion for, in the late 1970s, the Government ordered an investigation into long-term prospects for a second Severn road bridge to supplement the graceful structure opened by the Queen on 8 September 1966, just as BR once considered a second tunnel to meet traffic demands in the early 1950s when a post-war boom was in progress.

The Tunnel and all Bristol's main lines were included in the *Review of Main Line Electrification* published by BR and the Department of Transport in 1981. It also included Paddington–Taunton direct and on to Plymouth/Penzance); Paddington–Birmingham–Shrewsbury via Oxford; Bristol–Birmingham, direct and via Worcester and Kidderminster; Gloucester–South Wales, and Swindon–Gloucester.

Routes from distant places to Paddington—I have in mind those of the former Cathedrals Express from Hereford, and the Cornish Riviera through Cornwall and beside Channel shore at Dawlish—slice through more varied scenery than those from Bristol. But Bristol can historically claim the prototype. How many main lines in the world were so well engineered with the graciousness of viaducts like that across the Thames at Maidenhead, or the imposing dignity of tunnel entrances like Box?

The Midland Railway went to Bristol as well, piercing like a burning lance what the GWR regarded as inviolate territory, withering and destroying (although not immediately) the Broad Gauge, the very foundation of the Company. Not until Edwardian days more than half a century later did the GWR retaliate with a Bristol–Birmingham route with a distance handicap of ten miles, although the Severn Tunnel's completion in 1886 produced ever-increasing competition, including a North-to-West express service via Shrewsbury that was a tremendous boon to people travelling some of the longest distances in Britain as they moved between Scotland and the West of England.

The Tunnel also placed the London–South Wales main line on Bristol's doorstep, passing within sight of Temple Meads, and partly displacing the heavily-graded and twisting route via Gloucester and the Cotswolds. The Badminton Direct Line of 1903 took pressure off the congested London–Bristol main line and a few years later made Avonmouth one of the closest ports to the Capital.

Further relief for Bristol came when the Reading–Taunton direct line opened in 1906, putting the West Country twenty rail miles nearer London.

Another major GWR through route that pierced the region was that between Paddington and Birmingham (and Birkenhead). There was a host of secondary lines that did much to open up huge rural areas. A notable one was Worcester–Hereford, including a long and deep tunnel through the Malvern Hills, the countryside of Elgar who, his biographers tell us, travelled much by train. Yet steam never caught his imagination in the way that he remembered the carts that rumbled slowly through remote villages in his childhood. They inspired *The Wagon Passes* in his 'Nursery Suite'.

But the excitement of the Bristol–Birmingham did attract a writer who also lived near it. In *Brensham Village* John Moore wrote of the *Adam and Eve*, a pub close to a wayside station, which was not only a village pub but also a railwaymen's, portrayed the local gangers "not very well paid" and the engine drivers '. . . who were the little lords of the railway line, to whom Brensham was but a signal and a station and who thundered splendidly past in their fiery chariots.'

The Bristol–Birmingham route has two cities and a town which the railways have never managed to serve satisfactorily, including Gloucester, where the long platform of today is not appreciated by all passengers, and where through expresses have had to reverse since the closure of sharply-curved Eastgate. At Cheltenham, the former Lansdown station is not as close to the centre as many travellers would like it to be. The old railway networks of the two places were very different, as E.C.B. Thornton pointed out in *Railway World* in July and August 1963. Cheltenham was never a railway centre in the sense that it was a busy junction. Its character was different to that of Gloucester, where passengers changed trains and stations. The other unsatisfactory railway City is of course Worcester, where to help relieve a traffic-congested centre and improve passenger services, a Parkway at the junction of the Bristol, Birmingham and Oxford lines has been much discussed and criticised in recent years.

The Thames & Severn Region includes some notable cross-country and secondary lines born of rivalry. Soon after I began this book, Monty—Field Marshal Lord Montgomery—died in Spring 1976 and it occurred to me that he was perhaps the best customer of lines like the Didcot Newbury & Southampton and the Midland & South Western Junction, used to capacity as the D-Day supplies flowed south. But war service did not save them in peace and both have closed, the MSWJ noted as the Western Region's first major line economy.

Whole areas of middle England prospered from railways. Cheltenham and Gloucester and other towns quickly expanded and increased already well-founded prosperity; Stratford-upon-Avon's tourist trade boomed while country branches benefitted a class of tourist that some guide book writers openly despised. In his *Tourist's Guide to the Wye* (1892) G. Phillips Bevan stated:

The more lazily disposed, and that class which requires only a day to see that which ought to take a week, have the advantage of a railway . . . It may be as well to remind the tourist that refreshment rooms are few and far between, being only to be met with at Hereford and Builth Road above Builth.

As well as people, Bevan did not always like places:

Lydney is a dirty little town, depending upon iron and tinplate works, being also the shipping port for the Forest of Dean coals.

Railways turned Lydney into a boom town, for few areas were so well served by railways, which had displaced primitive horse tramroads.

More than a decade ago the steam locomotive which displaced the tramroads passed into retirement, if not quite history, and today the cheerfully-coloured HSTs are performing feats which the great steam expresses like the Bristolian and the Cheltenham Flyer could never have approached. Paddington–Bristol/South Wales HST services have pulled back passengers in their hundreds of thousands and another HST service North East/South West is in prospect for the Bristol division (which also has control of the Penzance HSTs). These trains do not carry fare supplements, as did their curtain-raisers: the 'Blue Pullman' diesel multiple units, which operated in the 1960s on the Bristol and Birmingham routes from Paddington and, for a shorter period, on the South Wales main line.

The fortunes of the railways in the Thames and Severn region have been influenced to some extent by the region's position on the fringe of several of the largest population areas in Britain: Greater London, the West Midlands, South Wales and (when swollen by summer holidaymakers) the West of England. Beeching and earlier economies have wiped out 'pockets' of highly individual local railways, notably those of the Forest of Dean and the neighbouring areas and in the lonely countryside of the Welsh Border.

Opening and other important dates are given, but it has been found impossible to include a detailed chronology of all but the Bristol area lines because the Region abounded in small lines with complicated histories. Only the more important tramroads and projected lines are mentioned. Spelling of place names–notably Stratford-upon-Avon–are given in the contemporary contexts of events. Local government re-organisation of 1974 introduced radical boundary changes, notably creation of the County of Avon, based on Bristol and Bath, and stretching along Severn shore from Western-super-Mare to north of Thornbury.

CHAPTER II

From London to Bristol

There is no better place from which to appreciate Brunel's visions and achievements almost a century and half ago than from the cab of án HST, which I reached through the engine compartment, having been issued with regulation ear-plugs—something I had not been given for footplate or cab visits before. The very fact that the London–Bristol main line with its generous curves and spacious junctions and cuttings had to be adapted so little for trains running at speeds Brunel could never have imagined is a fitting memorial, though far from being his only one.

Beside twisting Thames, straight and direct through the Vale of the White Horse, over wooded Cotswold flank and through the heart of elegant Bath, he drove a line that might have so easily ravaged the countryside, and yet with finely proportioned engineering works added, rather than distracted from it. He built a railway, something which Britain had to have to transform itself from a period of stagnation into one of industrial revolution and prosperity, with a minimum disturbance – a railway on which the still much-feared steam locomotive could work efficiently.

The opening of Bristol's first docks in 1809 coincided with an upsurge in trans-Atlantic trade and stimulated interest in the possibility of long-distance tramways, mainly to the English Channel, Plymouth being one suggested terminal. Nothing came of them and the City was still without a railway when the Liverpool & Manchester opened in 1830.

In that year, a builder who moved from a Gloucestershire village to Merseyside to develop New Brighton as a watering place wrote to his wife about joining him:

You will have to go to Bristol and book seats in the coach, but I would advise you to send the furniture to Dublin as there is better communication between Bristol and Dublin, than Bristol and Liverpool.

It was another eleven years before Bristol was on the railway map. Considering that steam railways were by then so well established, there was a lamentable lack of overall thinking and planning. Did promoters seriously believe that the London & Bristol line and the Bristol & Exeter Railway (Volume 1), would not be welded into a trunk route between London and the West of England? It seems they did, for the first half-mile or so of the B & E approached Temple Meads at absurd angles, with a platform line that could handle only a trickle of through traffic. What a fiasco, considering how successfully the London & Birmingham and Grand Junction had been brought together as a trunk route.

One of the first seriously considered schemes was that of the London & Bristol Rail Road Company of 1824 (which prompted plans for the Bristol Northern & Western Railway Company with lines to the Midlands and the West). Although it failed through lack of support, the L & BRRC was Bristol's first projected railway designed for locomotive operation. Among its directors was the road engineer, John Loudon McAdam, then surveyor to the Bristol Turnpike and many similar trusts. He became the railway's engineer and within a fortnight projected a route from Bristol through Mangotsfield, Wootton Bassett and the Vale of the White Horse. Regarded as too ambitious and costly the scheme died, but spurred creation of two local tramroads in 1828: the Bristol & Gloucestershire Railway, and the Avon & Gloucestershire, to serve coal mines around Mangotsfield.

BIRTH OF THE GREAT WESTERN RAILWAY

The opening of the Bristol & Gloucestershire tramroad on 6 August 1835 was followed only three weeks later by the London & Bristol Act of the Great Western Railway, the most momentous event in Bristol's railway history then or since. Routes had been under discussion for years, the Bristol & London Railway scheme of 1832 favouring one to serve several small intermediate towns: Bath,

Plate 1
Above: Broad gauge retreat. In the last days of the broad gauge, a Paddington–Plymouth train leaves Sonning Cutting in 1892. (*GWR*)

Plate 2
Below: Didcot was a major junction between the West of England and Oxford and Birmingham line. It is also the end of the four track section from London. Today the Didcot depot of the Great Western Society occupies the site of the motive power depot seen in the right background. (*L&GRP, courtesy David & Charles*)

Plate 3
Above: Commuter route terminal – Henley-on-Thames 1964. A diesel railcar seems almost lost amid long platforms of a dignified branch terminal much bigger than its 'neighbour' – Marlow. (*G. H. Platt*)

Plate 4
Below: The well-wooded branch terminus at Blenheim & Woodstock in the 1930s with a GWR push-pull train waiting to leave for the main line junction station at Kidlington. (*L&GRP, courtesy David & Charles*)

Bradford-on-Avon, Trowbridge, Hungerford, Newbury, Reading and Southall. Paddington was to be the terminus with branches to the projected London & Birmingham Railway and the City of London. Hopes of a 15 per cent return on capital were not sufficient to tempt investors.

But the survey stimulated four influential Bristol businessmen (who met in an office on what became the site of Temple Meads goods depot) to form a committee of leaders of the Corporation, the Society of Merchant Venturers, Bristol Dock Company, the Chamber of Commerce and the Bristol & Gloucestershire Railway. It first met in January 1833 and on 7 March appointed Brunel (at 27, half-way through his life) engineer. He was well-known locally as the docks engineer and had had a design accepted for a suspension bridge across the Avon Gorge at fashionable Clifton. He quickly surveyed, dismissing a route through Pewsey Vale in favour of what became a reality. Boards set up in Bristol and London first met on 19 August and the Great Western Railway came into being.

Fame was not established overnight. There was nothing *fait accompli* about the birth of the GWR. The *History of the Great Western Railway* (MacDermot, revised Clinker) where the full story is told, points out that some South of England investors were even apathetic to the Bristol scheme. The line was projected as likely to boost Ireland's agriculture and that of the West of England (ahead of the Bristol & Exeter). Cornish fish was to be shipped to Bristol and speedily sent to London.

By going through the Thames Valley and north of the Marlborough Downs, Brunel kept gradients virtually to nil for mile after mile, and provided an artery for opening-up much of the West and Wales, with branches to Oxford, Gloucester and South Wales, which could not have stemmed from a line through Pewsey Vale. Public support was lobbied over wide areas, including the West Country and Ireland.

It was to no avail, and Parliament's rejection of the GWR Bill gave the London & Southampton Railway promoters chance to press for a branch to Bristol from their own line at Basingstoke. This went before Parliament (with a revised London–Bristol line) restricted to Basingstoke–Bath, although the rest of a route had been surveyed. A House of Lords Committee heard evidence for forty days, a time when the London–Bristol supporters were never in a wilderness of confused thought. They included George Stephenson and Joseph Locke, whose reputation and experience outweighed that of opponents and the L & B got Royal Assent on 31 August 1835.

It was a purpose-design speed track to give Bristol the fastest and most direct link with the Capital. Small and pleasant market towns like Wallingford, Wantage and Faringdon were ignored and so the main line trains had a 40-mile clear run west of Reading.

The next momentous date was 29 October 1835 when the GWR directors adopted the Broad Gauge of 7ft 0¼in. Gauge stipulations had been deliberately excluded from the Act after Brunel persuaded the Lords that the question should remain open so it could be decided which was best for speed and passenger comfort.

Gracious viaducts (of which Wharncliffe and Maidenhead are examples) and ornately-mouthed tunnels matched the opulence of the gauge and spirit of the age. The first section from Paddington to a station about a mile short of Maidenhead opened 4 June 1838. The town was already expanding, with a population of 10,000 (which has since swelled five times).

The early coaches rode badly, locomotives were unreliable, but the GWR directors were consoled when they inspected the London & Birmingham, which began a restricted service over its entire length three weeks later, and found the track but little better.

The route was completed in nine sections, Twyford being reached 1839 and (after troublesome works in Sonning Cutting) Reading the following March. The town was given the first of several one-sided stations with up and down platforms end-to-end. Brunel designed them for 'one-sided' towns—towns which lay almost entirely on one side of the line. Brunel saw two advantages: passengers did not have to cross lines and non-stop expresses could pass clear of the platforms. But only one train at a time could enter the station and there were numerous conflicting traffic movements.

As construction progressed, five sections were opened in 1840, the last being to Hay Lane, four miles from Wootton Bassett, reached just before Christmas. It meant that construction carried out under the direction of the London board had been completed, for the 'boundary' between the directors' divisions was at Shrivenham. Bristol–Bath had been joined in August 1840, but the link-up through the Cotswolds took time; Box Tunnel was the longest yet attempted, being 786 yd longer than Kilsby Tunnel on the London & Birmingham, and only 308 yd short of two miles.

Four thousand navvies worked day and night at Box, where the GWR paid for the construction of a south aisle for their use in the small parish church. The grand portals of the tunnel were a public relations exercise (like the Euston Arch) to show the public that railways could be dignified—and were valuable. Besides Box Tunnel, a host of heavy engineering works were packed into the Chippenham –Bath section of some thirteen miles: long and deep cuttings, embankments, bridges, and work necessary to divert the Kennet & Avon Canal.

Only a single track had been completed through Box Tunnel when the London & Bristol main line opened, without public ceremony, on 30 June 1841, together with the Bristol & Exeter between Temple Meads and Bridgwater. The directors' special train ran through from Paddington. A string of stations lay between the cities, mostly serving growing towns at either end and villages in the middle section. The original stations were West Drayton, Slough, Maidenhead, Twyford, Reading, Pangbourne, Goring, Moulsford, Steventon, Faringdon Road (Challow), Shrivenham, Wootton Bassett Road, Chippenham, Corsham, Box, Bath and Keynsham. Two notable exclusions were Didcot and Swindon.

The village of Didcot began growing into an important junction with the Oxford branch opening in 1844. An imposing station had an all-over roof spanning four lines and five narrow platforms. It was burned down in 1885 and replaced by a more spacious one as further development was taking place: Didcot West curve opened 15 February 1886. This completed a triangular layout of which an avoiding line to carry Oxford line trains clear of the station was another strategic feature. As a mixed gauge line it was crucial to railway politics in the 1850s when a narrow gauge route was being established between north and south via Oxford (chapters IX and X). Today it provides a grandstand from which passengers can glimpse the depot of the Great Western Society.

SWINDON—HISTORY AT ITS HEART

Swindon was only a small village when it became a junction with completion of the Cheltenham & Great Western Union to Kemble (fourteen miles) and Cirencester, the day before the main line opening, and when the GWR chose it to be a railway town destined to be greater even than Crewe in the sense that it had carriage and wagon shops as well as locomotive works.

Recently the railway village which lay at its heart has been restored to set it apart in atmosphere from modern Swindon, so much like other industrial towns that have grown and changed. Like Crewe, Swindon is no longer purely a railway town, the car and other industries having moved in. In 1976, when it had a population of about 90,000, Swindon projected itself as 'the most successful expanding town in the country.' It invited firms to flee London in advertisements pegged to a 12 per cent rise in *rail* fares, the fourth in that year.

The Works has progressively shrunk. Some 14,000 people were on the payroll in the early 1920s. Today the total is 5,000. Swindon was once the only town in the Western Region where everyone went on holiday at the same time. In 1951, some 18,000 people used thirty trains to get to resorts.

Swindon was the GWR's home of passenger comforts. The General Stores kept a small stock of rugs and pillows for 'temporary requirements,' which included football excursions! Spiritual comfort was provided at a Wesleyan Chapel just outside the Works. Today, as a modest yet intensely interesting railway museum, this is a place where thousands worship bygone steam. The HSTs that race through the station are not Swindon products, a fact which underlines the tremendous change there has been since the end of steam. Output is diverse: in 1979 British Rail Engineering, Swindon, turned out its first-ever export locomotive, a Rolls Royce engined diesel shunter for Kenya. The Works has also modified the bodies of nearly 200 single-deck buses for the Bristol Omnibus Company (1973) and built forty-two bodies for export coaches under an order placed by the Gloucester Railway Carriage & Wagon Company (1974). That was fourteen years after it built its last steam locomotive for BR, No 92220 *Evening Star*. The hand-over ceremony on 18 March 1960 was soon followed by output of a *Western* class diesel-hydraulic locomotive.

MAIN LINE: ALMOST A CENTURY OF CHANGE

Between the Severn Tunnel opening in 1886 and the Badminton cut-off seventeen years later, the L & B line via Bath carried increased traffic, partly because the tunnel provided a direct line between South Wales and Portsmouth and Southampton at a time when Britain's large navy and a fleet of large Trans-Atlantic liners were burning prodigious amounts of coal. The traffic caused congestion in the Bristol area and improvements there included the creation of a triangular junction just outside Temple Meads by the addition of a spur between North Somerset (originally Feeder Bridge) Junction and Dr Day's Bridge, opened only a few months earlier. Nearby, Bristol East Depot was opened, a marshalling yard of more than thirty dead-end sidings squeezed between the east bank of the Avon and a hillside. One of the L & B tunnels of 330yd was opened-out for a shunting neck. The yard (1890–1967) caused operating problems because the Up and Down sections lay on either side of the main line, making transfer work slow and complicated.

The Bristol main line has seen many changes in recent years, carrying ever-increasing commuter traffic, handled locally by diesel multiple-units and at greater distances like Didcot and Swindon, by HST and other Inter-City trains. The Thames Valley section between Reading and Didcot gained extra importance following concentration of South Coast—Midlands traffic via Reading West after the closure of cross-country lines like the Didcot, Newbury & Southampton.

West of Didcot, the Bristol line has lost not only most of its branches, but its wayside stations, closed 1964–65, leaving only Swindon and Chippenham between Didcot and Bath. But the innate, glorious character of the line remains and so does much of its original setting, for west of Reading the countryside of woods and green fields has been little blemished (with the exception of Didcot power station) since the navvies first changed the landscape.

The deep and abiding loss has been of steam, for this was a main line on which the great locomotives—Kings, Castles, Saints and Stars among them—were able to show off their power. The GWR celebrated its centenary in 1935 by introducing The Bristolian, which maintained 1 hour 45 minutes schedules up to 1939; these timings did not return until 1954 for a final burst of steam glory

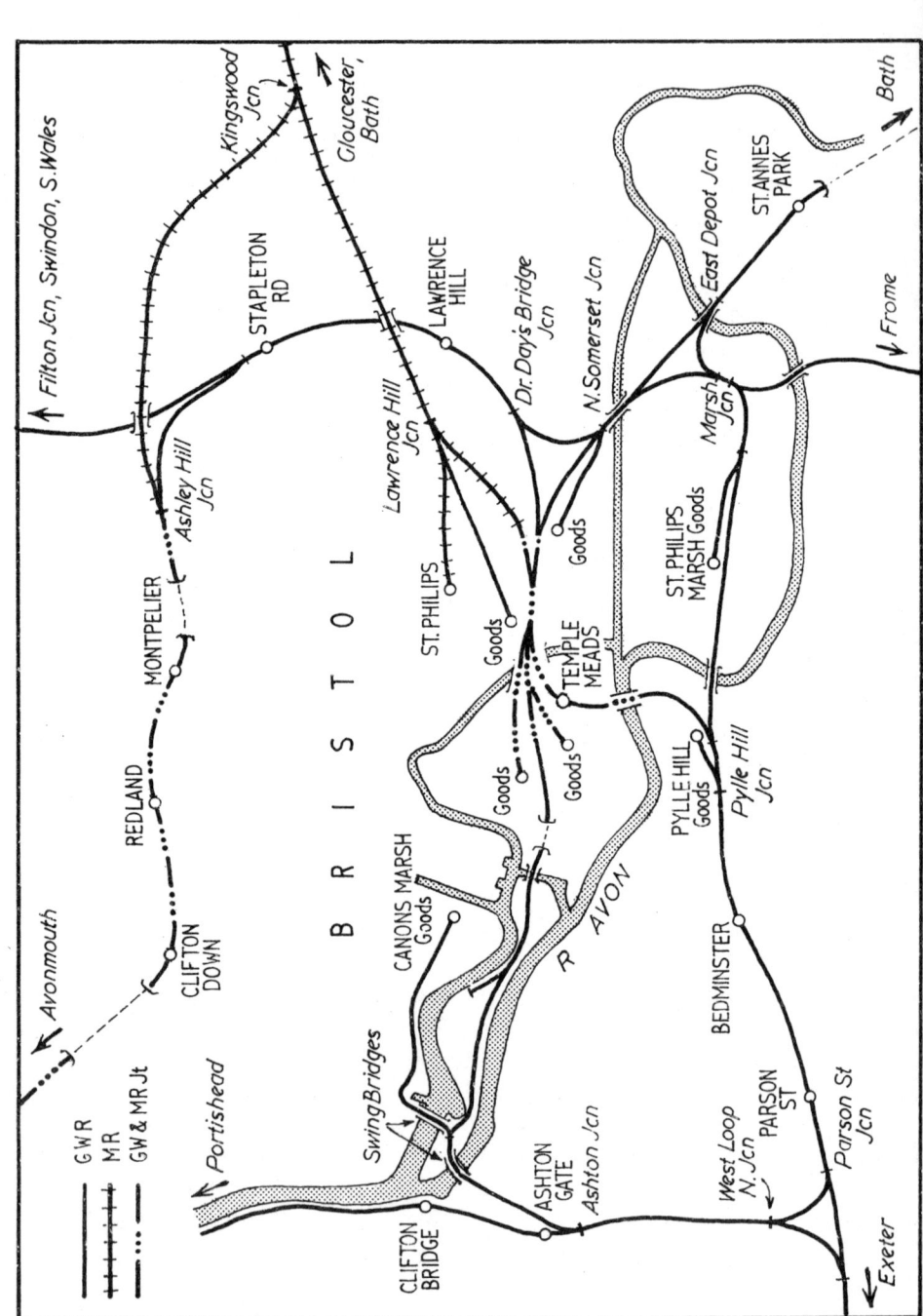

until diesel locomotives knocked five minutes off timings and held them until 1976 when HSTs brought the journey time, with two stops, to just over $1\frac{1}{2}$ hours between Paddington and Temple Meads. Bristol Parkway can be reached almost fifteen minutes quicker.

The speed potential is far from exhausted. In 1975 a prototype APT ran at 151mph for five miles nearing Goring-on-Thames.

TEMPLE MEADS JOINT STATION

The ambience of the old Great Western is likely to become stronger at Bristol than anywhere else as a result of BR's lease of the original station buildings, now overshadowed by office blocks, to a preservation trust for development into a Brunel engineering museum and other projects. The train shed, which covered five broad gauge tracks under its 72ft hammerbeam roof, and has been used as a car park since 1965, is to be restored at a cost estimated in 1981 at £1,500,000. Underneath, where stables were converted some years ago, the 130-seat Brunel Theatre flourishes with civic support.

Growth of the original station took years because of a profusion of plans to build one closer to the city centre on the north side of the Floating Harbour. Shortage of money and local apathy led to the downfall of several early schemes and three companies (reduced to two after the GWR absorbed the B & E) got powers for expansion which confirmed Temple Meads as the city's main station. Work between 1865–78 created the magnificent stone turret frontage of today, and behind got rid of the awkward junction between the GWR and the B & E. Later, two island platforms were squeezed in on land freed when the broad gauge was lifted in 1892.

A development that same year which eased station congestion was the Bristol Relief Line, opened 10 April, running just over a mile between the London line at East Depot and the Exeter line at Pylle Hill. It joined the North Somerset branch and from 1910 provided access to the new St Philip's Marsh locomotive sheds. Its operating drawback was that it could not be used by Midland expresses on the Birmingham main line. They had always to run through Temple Meads to head west.

The misery that holidaymakers suffered through delays, which often added hours to already long journeys, was reduced after the rebuilding of Temple Meads in the mid-1930s, when by further bridging the Floating Harbour platforms were lengthened and increased, partly by using scissors cross-overs, from nine to fifteen. The station was almost trebled from its original size and associated work (mainly to relieve unemployment), included main line quadrupling between South Wales and Portishead Junctions and the rebuilding of Horfield, Ashley Hill, Bedminster and Parson Street stations.

Temple Meads itself remained a place where virtually every train stopped. GWR Working Appendices of the mid-1930s noted that it

> must be regarded by Enginemen as a Terminal Station, and they must have their Trains under complete control, relying upon their Hand Brakes only, and be prepared to stop at any part of the Platform as directed by the Inspector in charge.

After World War II Temple Meads once more became a holiday weekend bottleneck. Controllers working inside another Brunel building at the station approach, did what they could. During a visit a decade later, I still heard tales of many serviceable steam locomotives suffering 'unavoidable delays between Thursday and Saturday' at Bristol. Locomotives for Saturday extras to the West were always in demand.

Temple Meads was never able to cope with the holiday rush until the problem was solved for the railways by the motorway builders. The station assumed its present form in 1966 when the two original platforms were divided and Nos 12 & 13, 14 & 15 closed. These were partly in the Brunel train shed used by LMS express and local services, including those to Avonmouth. After track lifting, Bristol Power Box was built at the mouth of the train shed, tucked away behind the main station wall.

Temple Meads remained Bristol's undisputed main station because of the failure of a host of schemes. Probably the most hair-brained was for a direct line to South Wales, suggested in 1854 from a central station in Queen Square, through the Avon Gorge to New Passage and across the estuary on a floating bridge. The promoters seem to have ignored the Bristol Channel's reputation for having some of the fastest-running tides in Britain.

A central station in Queen Square featured in a Bristol & Clifton

Railway proposal of 1861 designed to link up with the projected and strongly favoured, but isolated, route of the Bristol Port & Pier Company, and also to serve Bristol Docks. Support was impressive: 5,000 people signed a petition for a central station and the scheme was supported by Chambers of Commerce in Birmingham, Sheffield and Leeds, as well as Bristol itself. The GWR agreed to provide half the £250,000 capital, but because local councillors and organisations feared the line would take trade away from the Docks, a Commons Committee rejected the scheme in 1862. A rival scheme of the same year was for a dock tramway and central station in Stone Bridge, advanced by the Bristol Central Railway & Terminus Company, which publicly blamed the jealousy of the big companies for its downfall. Bristol Corporation opposition destroyed the Bristol Railways Junction scheme of 1862 to link up with the Avonmouth line.

A scheme that actually got Royal Assent was the Bristol Port Extension Railway, 1864, to connect the Avonmouth line with the main lines in Bristol and the Docks. Its central station was to be in Christmas Street. It failed to attract investors, mainly because in 1866 the GWR got authority for the Bristol Harbour Lines. They provided for everything—except a central station.

RIVAL STATIONS: RIVAL ROUTES

A variety of central station schemes was also projected by companies attempting to break the GWR monopoly between Bristol, London, and the West. They centred mainly on several quite well-founded attempts to link Bristol with the London & South Western main line. That company projected a branch from Gillingham via Frome, but it was rejected by Parliament in 1862 together with a GWR attempt to reach Southampton and to build in Bristol itself a Clifton branch with a central station in Queen Square. Even if its line had been built, the L & SWR would have been hopelessly handicapped with a route between London and Bristol of 147 miles—twenty-seven miles longer than that of the GWR.

More promisingly competitive at 130 miles was the Bristol & London & South Western (Junction) Railway, promoted by 150

Bristol businessmen in 1883. They contemplated doubling the Bristol & North Somerset (Vol 1) and reaching the L & SWR at Grateley, near Andover. After objections by the Midland Railway, the scheme was reduced to a short branch to the Somerset & Dorset at Wellow to allow the Midland to take L & SWR expresses into Temple Meads via Bath and Mangotsfield. The outcome was not a branch, but a ten-year GWR—L & SWR agreement not to invade rival territory.

Construction of the Royal Edward Dock at Avonmouth inspired the Bristol London & South Counties Railway of 1902. It was to have its own stations at Bristol and Bath, and by crossing Salisbury Plain put Waterloo slightly nearer Bristol than Paddington. Bristol Corporation promised to raise £100,000, but £6,000,000 was needed and a House of Commons Committee rejected the scheme, stating that there was no evidence to show where the money was to come from—except, said the chairman, the energetic expression of the pious opinions of Bristol witnesses that the money would be forthcoming.

BRANCH LINES: LONDON—BRISTOL

Rival branches opened to WINDSOR in 1849 were the GWR's 2½-mile line from Slough to what became Windsor & Eton Central (8 October), and the L & SWR's to Windsor & Eton Riverside (Volume 2). Bitter opposition forced the GWR to drop a branch from its London & Bristol Act of 1833. 'No public good whatever could possibly come from such an undertaking,' wrote the Provost of Eton. The branches might have been joined but for local opposition, which also defeated 1846 schemes for the London & Windsor Railway and the Windsor Slough & Staines Atmospheric. The GWR ran slip coaches between Paddington and Windsor, while today Windsor can be reached in twenty-three minutes for the twenty-two mile journey via Slough, served by High Speed Trains.

A branch which the GWR inherited was Bourne End—MARLOW (2¾ miles), which had been built by the wealthy for the wealthy. The Great Marlow Railway Company of 1868 which took its name from the small riverside town, then called Great Marlow, was owned by local businessmen; two Members of Parliament were among the original seven directors. It opened on 28 June 1873 and the company was absorbed by the GWR in 1897. Immediately an extension to the Henley branch was proposed, closely following the Thames and twice crossing it. The extension was conceived as double track, with the Marlow and Henley branches upgraded to form a grand circular line through the Thames Valley.

The Bill went to Parliament, only to be dropped in the face of opposition from local authorities and landowners. The Marlow branch survives on the 'Western Commuteroute' and its future is guarded by the Marlow/Maidenhead Railway Passengers' Association, formed in 1972 to fight a closure threat. It stated that the associated Bourne End—High Wycombe section of the Wycombe Railway had been closed 'quite suddenly with little time for organised public protest. Local rail users were determined that this should not happen with the rest of the branch.'

A far more intensive commuter service is carried by the Twyford —HENLEY-ON-THAMES branch, although through services to and from Paddington were withdrawn in the winter of 1975. The GWR got authorisation for the $4\frac{1}{2}$-mile branch in 1847, but did not open it for ten years, on 1 June 1857. Broad gauge, it was converted to narrow in 1876, doubled in 1897 and closed to goods in 1964. For years it was used to capacity during Regatta week. In 1919 the Working Notice of Special Trains for the four-day period ran to thirty pages of small type. It stipulated:

All coal, Mineral and unimportant Goods Traffic for Twyford and Henley-on-Thames must be held back at Reading West Junction, or as arranged by Control. The 7.50pm Milk Train from Chippenham to be kept clear of Regatta Specials. Booking arrangements: All Available Windows at Paddington to be open from 9am until 1.0pm.

The WALLINGFORD branch was another that was once planned to cross the Thames, but never did. Authorised in 1864 as the first narrow gauge off the London—Bristol route, the Wallingford & Watlington Railway (nine miles) never went beyond the market town. Until the branch opened on 2 July 1866, Wallingford's railhead was Wallingford Road. It was then renamed Moulsford, and branch trains ran three-quarters of a mile alongside the main line. The GWR absorbed the independent company in 1872—the year the Watlington branch opened (Chapter IX). When the London–Bristol route was quadrupled in the 1890s Moulsford station was replaced by one at Cholsey, and the branch journeys were cut back by nearly three-quarters of a mile. Passenger trains stopped in 1959 and goods in 1965, when the line was cut back from 3¼ to 2¼ miles, to Associated British Maltsters mill. The branch closed completely on 1 June 1981.

The inhabitants of WANTAGE overcame their handicap of being left isolated from the London–Bristol main line, only 2½ miles away, by forming a local company to build a half-tramway, half-railway to Wantage Road station. The Wantage Tramway Company formed in 1873 began goods operation with horse trams 1 October 1875 (passengers 10 October), and with Parliamentary authority used steam tram engines soon afterwards. Passenger trams were withdrawn in 1925 but goods trains continued profitably until 1946, when there was an enthusiastic GWR response to a *Times* reader's suggestion that the 0–4–0 well tank *Shannon* deserved preservation in recognition of its long service to local people.

The 3½-mile Uffington–FARINGDON branch was the result of private initiative to open-up a large farming area. Broad gauge from its opening 1 June 1864 until 1878, it was absorbed into the GWR in 1886. Its passenger service, which survived until December 1951, included a 9.45am to Faringdon which for years ran only on the first Tuesday of each month. Goods trains were withdrawn in July 1963, when the branch closed eleven months short of its centenary.

A private company more ambitious territorially was the Swindon & Highworth Light Railway Company, formed 1875 when there were hopes of a light railway from Swindon to the East Gloucestershire Railway's Fairford branch at Lechlade The idea was soon dropped and the 5½-mile HIGHWORTH branch served industry near Swindon and the little town, beloved by Sir John Betjemen in his book *First and Last Loves*. The S & H was soon in financial difficulties and the GWR reluctantly took over nine months before the line opened 9 May 1883. Soon GWR workmen's trains began running to Swindon. With others from Purton and Wootton Bassett, they were an institution for more than fifty years. The Highworth trains continued long after scheduled passenger trains were with-drawn in 1953, nine years before complete closure.

Early in the evening of 10 September 1951, stationmaster P. J. Wood of MALMESBURY prepared extra tickets because 200 passengers were expected to catch the last regular passenger train, the 7.25pm to Little Somerford. Only twenty-five travelled, yet that was five times the usual number. The 6½-mile Malmesbury Railway was sanctioned in 1872 from Dauntsey to Malmesbury—'a quiet, unprogressive town stagnant for a century', said a guide-book thirty years later. The GWR subscribed half the capital of the private company, which it absorbed three years after the line opened on 17 December 1877. The branch suffered an unusual amputation on 17 July 1933 when its junction was switched from the London & Bristol at Dauntsey to the Badminton route at Little Somerford, where a spur was built and the intervening 2¾ miles between the main lines closed completely. The truncated branch, finally closed in November 1962, had been the outcome of ambitious plans by the Midland Railway to reach Salisbury from the Bristol–Birmingham main line by upgrading the Stonehouse–Nailsworth branch. They were forgotten once the Midland and GWR reached territorial agreement.

An arbitrator decided that because of it, the Midland could not work the Wilts & Gloucestershire Railway, authorised in 1864 from Christian Malford on the London & Bristol, four miles from Chip-penham, to Nailsworth: a proposed mixed gauge route of twenty-three miles. The North & South Wiltshire Junction, authorised the following year from Christian Malford to the Berks & Hants at Beechingstoke (seventeen miles) also quickly faded away. Had the Midland achieved its ambitions, the route would have duplicated the Cheltenham & Great Western Union, and served Malmesbury, Tetbury and Nailsworth, and run just east of CALNE, where the local branch was to be extended a quarter of a mile to join it. The 5¼-mile single line from the London & Bristol at Chippenham was opened broad gauge on 3 November 1863 by a private company. 'The rail from Chippenham to Calne is pretty but calls for no special remark,' stated *Baddeley*. For years this was a busy branch. Special passenger train vans worked out of Calne made the town a household name throughout Britain with their distinctive headboards: *Harris (Calne) Wiltshire Sausages—Calne to* . . . During World War II, thousands of RAF personnel travelled the branch to reach large radio training schools at Yatesbury and Compton Bassett. Afterwards decline began and Dr Beeching recommended closure, but the freight service withdrawn in November 1964 was outlived, unusually, by that for passengers, which survived until the following September.

CHAPTER III

Bristol and Avonmouth

Greater Bristol's rail network includes groups of lines that can be loosely sub-divided: the Bristol dock lines, GWR-connected, which never carried passengers; three routes to Avonmouth carrying freight to and from the GWR and Midland main lines and local passenger services, mainly between Bristol and Avonmouth; and dock lines at Avonmouth.

BRISTOL HARBOUR LINES

Brunel's genius was far from being confined to railways. Its variety was well demonstrated at the time when his main line was approaching Bristol, for it was then possible for a traveller from London to New York to catch a train on his line, at least part way to the docks at Bristol, which he improved. From there he could catch a ship that Brunel had largely designed, the 1,340-ton *Great Western*, which from the start of its trans-Atlantic sailings in 1838 carried nearly 6,000 passengers in six years. One hundred thousand people were reputed to have watched its first return sailing from New York to Bristol where a correspondent of *The Times* saw its arrival, caught a coach (there were then about twenty a day to London) as far as Maidenhead and transferred to the Great Western Railway, opened only a few days earlier.

Although Avonmouth docks overshadowed Bristol's from the moment they were built, the City docks were busy for decades. Yet they were not rail-connected until thirty years after the London main line opened. The first successful initiative came from the

BRISTOL HARBOUR RAILWAY COMPANY, incorporated in 1866 with the financial backing of the GWR, the Bristol & Exeter and the Corporation. Operated from opening on 11 March 1872 by a Joint Committee, the line passed just north of Temple Meads, crossed Victoria Street and ran through the burial ground of the tall-spired church of St Mary Redcliffe. It crossed Bathurst Basin on a steam-operated bascule bridge, of which relics are preserved in Bristol City Museum. The branch length remained at just under a mile until 1906 when it was extended ¾-mile alongside the New Cut of the Avon to join an important dock line opened the same day, 4 October: the CANON'S MARSH branch, serving a spacious goods station, closer than any other to the City centre. It took pressure off Bristol's congested network by enabling West of England freight trains to reach the docks without circling Temple Meads. It was the only line to the north berths of the Floating Harbour.

The branch, which ran from Ashton Junction on the Portishead branch of 1867 (Vol. 1), crossed Ashton New Cut on another unusual bridge—a hydraulically-operated two-decker, with a road on top. As the waterway fell into disuse, Bristol Corporation obtained powers in the 1950s to 'fix it' by removing machinery. The branch crossed the Floating Harbour on a swing bridge. The close proximity of the docks to residential, commercial (and spiritual) districts of the City were reflected in Service Time Tables, which in the 1930s, (not an age conscious of noise pollution), clearly laid down that 'Drivers of engines passing over the Harbour Line are required to cross Victoria Street and Pile Street Bridges at such speed, and with such caution in the use of their whistles, as will prevent any unnecessary noise.'

Even more specific was a section devoted to working at Canon's Marsh on Sundays: 'In accordance with an undertaking given by the Company no Train or Engine must be run between Lower College Green Avenue and the road known as the "Butts", nor must any shunting or moving of any Train or trucks be carried out during the hours of service at the Cathedral on Sundays in a manner which is likely to be audible in the Cathedral.'

Besides quays, the dock lines serve the sidings of private firms and a large gasworks close to Canon's Marsh depot. Canon's Marsh Goods Depot closed in 1965, but part of the branch still feeds a coal concentration depot at Wapping Wharf.

Plate 5
Above: Eccentric splendour! Wantage train leaving Wantage Road on 11 September 1920. Locomotive No 7. (*Ken Nunn collection, courtesy Locomotive Club of Great Britain*)
Plate 6
Below: Showpiece dignity. Veterans in the Great Western Railway Museum, Swindon. Small, intimate, delightful atmosphere. (*Swindon Libraries*)

Plate 7
Above: Bristol suburban. A local train for Temple Meads is about to leave
Severn Beach in March 1950. The signals in the background controlled the
level crossing on the branch that was then open to Pilning. (*Real Photographs*)
Plate 8
Below: Dawn of the HST age. Bristol Parkway in original 'primitive' form with
open footbridge and platform fencing, on 28 March 1972. (*Western Region*)

The first major step towards Avonmouth becoming a port was the authorisation of a deep water pier in 1862. This was the year following the formation of the BRISTOL PORT RAILWAY & PIER COMPANY by city industrialists and councillors. It had headquarters in Westminster. Yet it was nothing more than a 5½-mile line, single and totally isolated, from a terminus squeezed between a sheer rock face and river almost underneath Brunel's suspension bridge, to Avonmouth, where a deep water pier was also authorised in 1862. The railway opened 6 March 1865 and there was talk of an extension through the Avon Gorge towards Temple Meads from the terminus, whose original name of Clifton was changed to Hotwells, (by which it is best remembered today), in 1891.

Three large companies—the Bristol & Exeter, the Bristol & South Wales Union and the Midland—were to subscribe and appoint directors, and although the plan was approved by a House of Commons Committee in 1864 (together with a Bill for developing docks at Avonmouth), local support was far from unanimous. Controversy raged for years. After four years of it, the editor of the *Bristol Times* declared: 'People have often ere now said to local editors, in a spirit of condolence: "When these docks and railways, the so-called progress and anti-progress dispute, are finished, what will you all have to write about?" ' Plenty as it happened. The extension was soon abandoned and the Port Railway was used purely by passengers, and their numbers quickly dropped following the failure of a scheme for a pleasure resort at Avonmouth for 100,000 visitors a year.

CLIFTON EXTENSION RAILWAY

To reach Avonmouth direct from Central Bristol and free the Port Railway from isolation meant extensive tunnelling and the BPR & P (Clifton Extension) of 1867 provided for about a third of the 3¼-mile line to run underground, notably at Clifton, where the Down was pierced by a 1,738yd bore. The line ran from Sneyd Park, beside the Avon, to Ashley Hill Junction, connecting with a 1½-mile purely Midland spur from the Bristol & Gloucester at Kingswood Junction. Within sight of Ashley Hill, it crossed the Bristol & South Wales Union, from which the GWR built a connecting spur from Narroways Hill Junction, abolished when the Avonmouth lines were extended south parallel to the B & SWU in 1888, (two years after the Severn Tunnel opening), as far as Stapleton Road, where the station was doubled to four platforms.

The Pier Railway Company never paid a dividend and was in a Receiver's hands from 1869. With financial resources of company and contractor exhausted by tunnelling, the very much incomplete line was transferred to the GWR and Midland in 1871. The first trains—GWR and Midland via Kingswood Junction—reached Clifton Down (Whiteladies Road) via Ashley Hill 1 October 1874 but High Court wrangles prevented them reaching Avonmouth Dock until 22 February 1877.

On that day, 15,000 people attended great celebrations to mark the opening of the Dock built by the Bristol Port & Channel Company; the railway received little praise, for it was not built yet to passenger standards. Those services did not begin until 1885, their introduction coinciding with completion of a new joint Dock station at Avonmouth. The GWR & Midland Railways Joint Committee ran the Extension line from 1894.

As trade, deep-sea and coastal, grew, both companies recognised the need for control of the line and they bought it in 1890 for £97,500—and spent a lot more in the next seventeen years carrying out improvements. Mainly they involved heavy engineering work necessary to double the riverside stretch between Sneyd Park and Avonmouth, originally built to almost primitive standards.

World War I brought heavy traffic, but marked decline afterwards allowed abandonment of the Sneyd Park–Hotwells section in 1922 to make way for the present Bristol–Avonmouth main road, the Portway. Traces of Hotwells station and the short tunnels remain.

RAIL AND DOCK CONFLICT

Plans for a much larger dock at Avonmouth put forward in late Victorian years when the British Empire was on the crest of a trade boom led to conflict between rail and dock interests—the very parties who had acted so much in unison in the original project. Avonmouth became part of Bristol in 1895 and two years later earnest discussions began about financing what became the Royal Edward Dock at Avonmouth, estimated to cost £2,000,000. Bristol

Corporation, which hoped for GWR and Midland support, got a dusty reply in 1897 from the GWR chairman, Viscount Emlyn. He did not think trade, present or prospective, justified the Corporation going ahead without an assurance of new Atlantic trade.

The following year the GWR General Manager, J. L. Wilkinson, returned to the attack. He said that the Company was spending over £500,000 at Bristol, largely at the desire of the Corporation to increase trade at the City docks: extending Avonmouth would reduce that trade. Bristol Docks Committee continued to hope that the new Avonmouth dock would be a railway one, subsidised by the GWR, which would offer low rates for through goods. Members considered existing rates too high. Angrily, Bristol City Council 'declined to acknowledge the right of the directors of the Great Western Railway to dictate or control the policy of the Corporation.'

Two years later, when tempers had cooled, the Corporation's dock Bill went through Parliament without railway opposition, receiving Royal Assent on 17 August 1901, early in the Edwardian era. It was estimated to cost £1¾ million, including a comparatively small sum (£44,000) for a new Avonmouth passenger station, dock sidings and the diversion of exisiting lines.

One reason for the GWR's reluctance to make heavy outlay at Avonmouth was that it saw more potential for Fishguard, as a port for both Irish traffic and Trans-Atlantic liners. Competition for the Atlantic traffic was intense but after the railway reached Fishguard the liners arrived, Cunard switching the *Mauretania*, then the largest and fastest liner afloat, in 1909.

FEEDER ROUTES

Once Avonmouth dock construction began, new feeder routes were needed and the GWR rather than the Midland provided two that ran across flat ground lying between Bristol's northern outskirtsand the Severn Estuary. They were without heavy gradients that made the Clifton line difficult and costly to operate.

The 7¾-mile single AVONMOUTH–PILNING line, which had been authorised in 1890 and opened 5 February 1900 was adapted rather than built, 1½ miles near Avonmouth being moved further inland to make way for the dock; a new junction, St Andrew's, was

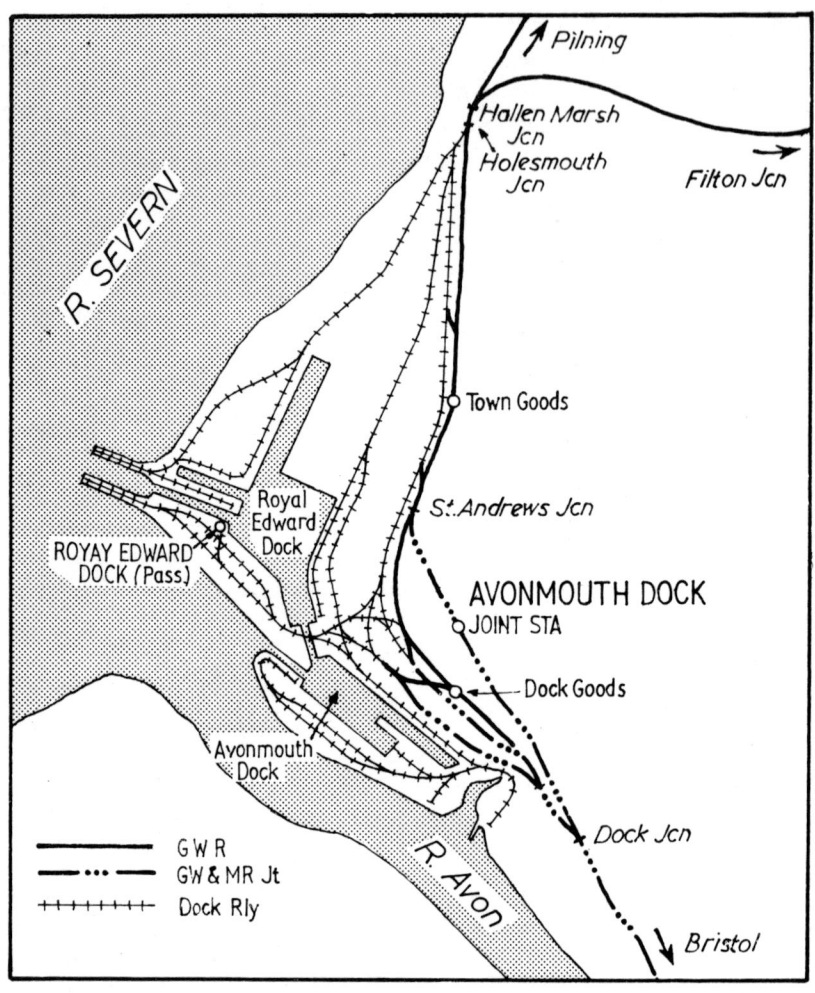

created with the Clifton route at Avonmouth. The line followed the Severn shore, used $1\frac{3}{4}$ miles of the B & SWU route to New Passage, closed since the Tunnel opened, and crossed the eastern portal of the Tunnel itself. Opened originally to goods, passenger services were not in operation until 1928 when Bristol–Avonmouth local trains were extended via a new station at Severn Beach, to Pilning, where a halt was established at Low Level, alongside that on the main line.

The GWR's main route to Avonmouth was the Stoke Gifford

direct line opened 9 May 1910, which joined the Pilning line at Holesmouth Junction, (which became Hallen Marsh Junction when the line was doubled in 1917). Not only did it put the Port within 120 miles of London, but also provided direct access to the GWR West Midlands route via Honeybourne. A useful roundabout route was developed via Filton, trains using the Joint Dock Station, which the GWR and MR had opened at Avonmouth in 1907. It is a line of varied fortunes: Filton Halt closed in 1915 only to re-open as North Filton Platform ten years later to serve the aircraft works. The line was doubled to meet wartime traffic demands in 1917 and singled again in 1966. Its traffic potential was increased in 1971 by completion of a north-to-west spur at Filton on earthworks unused since the line opened.

The Severnside network was completed by the AVONMOUTH LIGHT RAILWAY, which could easily have been mistaken for a siding, which it virtually was in all but name. Promoted privately in 1893 from the Port & Pier line near Dock station to Holesmouth Junction, it was to serve local works. The first section did not open until 1908, and when the company was purchased by the LMS and GWR under a Light Railway Order of 1927 it was still only about half its authorised length of two miles. Its long and slow moving history is told (with much else) in *Lines to Avonmouth*, by Mike Vincent.

The Monday–Saturday service by the Temple Meads–Severn Beach diesel multiple-units is a remnant of the much threatened, much curtailed services which the GWR and Midland both maintained to Avonmouth, the Midland running some through to Bath.

The GWR's Avonmouth services included the Filton roundabout, withdrawn in 1964, although North Filton Platform continued to be used by unadvertised workmen's services, and another roundabout via Pilning (Low Level), which also ceased in 1964. This was a year after the closure of an associated short branch to the Pumping Station close to the Severn Tunnel, and four years ahead of Severn Beach–Pilning complete closure.

The Severn Beach line provides access to ICI's large Severnside Works, opened in summer 1962. Also rail-connected are the Commonwealth Smelting Works at Avonmouth and the neighbouring Fisons fertiliser factory.

Only one rail approach remains to Avonmouth Docks—from

Holesmouth Junction to the Royal Edward Yard, worked by Port of Bristol Authority locomotives which, incidentally, cross BR lines to the smelting works. Dock operating was complicated. As GWR Service Timetable Appendices showed:

'The Royal Edward Dock Yard has been designed with the intention of obviating, as far as possible, obstructions being caused by Engines of different ownerships working over one and the same set of rails and separate Sidings with running Lines have been provided for this Company's traffic.'

'Clifton Down is the recognised transfer point for traffic from the Royal Edward and Avonmouth Docks (Bristol Corporation Railway), Avonmouth Dock Joint Station, Shirehampton and Sea Mills to the LM & S Railway (Midland Section) or beyond, and vice versa, and such traffic must not be allowed to pass via Stapleton Road and Temple Meads without the authority of the GW Divisional Superintendent.'

Of yards and sidings in the Bristol area, only about a quarter belonged to the Midland, but traffic was intense for years.

Dock lines and depots have closed as docks have closed. In a wider sphere, access routes have been singled or closed as shipping has declined or switched to container freighters until today, when Avonmouth, partly dominated by the hump of the M5 motorway bridge projects itself as a 'Motorway Port.' Its third generation dock, the £38,000,000 Royal Portbury, for ships of up to 70,000 tonnes, opened 12 April 1978, lies close to the Portishead branch. It is not rail-connected. Transhipment containers are taken by road to and from Kingsland Road Freight depot in central Bristol. In 1980 the Portishead branch was under threat of closure, BR stating that it could not justify on traffic grounds, expenditure of £1,000,000 necessary to retain it.

There was ever increasing traffic for the railways to handle. Two more arms of the Royal Edward Dock were opened in the 1920s. Imports reaching the Avonmouth, City and Portishead docks in 1937 totalled 2,800,000 tons, including 6,000,000 bunches of bananas. Petroleum and its products totalled more than 750,000 tons. Export traffic was small by comparison: 55,000 tons. The Royal Edward Passenger station, with its 'customs examination and waiting rooms, telegraph office, buffet etc' handled passengers on the weekly banana boats, cargo liners to the Far East and an increasing number of cruise liners. Again, the *Official Handbook to the Port of Bristol*: 'Special trains connect with arriving and departing liners, London or Birmingham being only a two hours' journey.'

A 'railway' with both stations close beside Brunel's suspension bridge like Hotwells, was the CLIFTON ROCKS RAILWAY, built 1891–3 by the publisher Sir George Newnes. He got authority from the Merchant Venturers' Society three years earlier to excavate a steeply-graded tunnel for four rope-worked tracks behind the Cliff face. It was a success for many years, but was closed in 1934 by the Bristol Tramways & Carriage Company, which had taken it over in 1912. Station remains can be seen at the top and bottom of the Gorge.

CHAPTER IV

From Bristol to Birmingham

Nothing can beat the appearance of the Midland "Diner" to Birmingham, Leeds and Bradford, and the Great Western Plymouth, Torquay and London "Corridor" express leaving Bristol side by side about 2p.m

Such was the scene that caught the imagination of Baddeley in his 1902 *Guide to Bath and Bristol and Forty Miles Round,* obviously aware that the Midland was a very different railway from the Great Western. The distinctive atmosphere of the Midland's Birmingham main line lingered after grouping when the LMS added its own contribution with red locomotives and coaches—often far more comfortable than those of the GWR because the third-class seating was three, rather than a four a side—a complication that bedevilled booking clerks long past Nationalisation.

Co-editor David St J. Thomas recalled:

Coming up from the west on a smooth Great Western track, changing to the Midland always meant dirtier trains and trains that *seemed* to be going faster, trains that had to perform feats such as often pulling up twice at Cheltenham, getting the banker attached ready for the Lickey incline or adjusting the brakes for the descent, and an extraordinary entry to Birmingham.

[The reference was to the roof-topping route of the Birmingham West Suburban Railway via Bourneville—Volume 7.]

The Bristol–Birmingham has had a far more chequered and interesting career than the London & Bristol, not least because of original gauge problems.

BROAD GAUGE CHALLENGE AND DEFEAT

On Monday, May 22, the narrow gauge line between Gloucester and Bristol was opened to the public. Passengers will be able to proceed direct from the north to Bristol without change of carriage, while some 10 to 15 minutes time formerly occupied in shifting passengers' luggage will be saved. A considerable saving will be affected by the company in the goods department, as every parcel and package conveyed from the north to the south had to be moved from a narrow gauge to a broad gauge truck.

The change of gauge chaos *Herapath's Journal* of 27 May 1854 reported as ended, reflected the tremendous volume of traffic that flowed along the Bristol–Birmingham route from its completion ten years earlier. The GWR suffered its first broad gauge defeat at Gloucester because of the guile of the Birmingham and Gloucester's goods manager, J. D. Payne, who discredited Brunel's assertions that change of gauge would cause few problems. Payne proved his point by arranging for two goods trains already dealt with in the tranship shed to be unloaded again during a visit of the Parliamentary Gauge Commissioners.

Before the railway age, the fastest means of carrying goods between Bristol and the Midlands was by a string of canals, of which the widest was Telford's Gloucester–Berkeley of 1827, which made Gloucester an 'inland port'.

Despite break of gauge, the Bristol–Birmingham railway extended the national network from Newcastle-on-Tyne to Exeter and enabled heavy goods from Birmingham, 'city of a thousand trades' to be quickly sent for shipment at Bristol. It also gave Gloucester and Cheltenham a second railway outlet, while branches that were soon to be developed opened up much of the Severn Valley. Gloucester had a population of about 10,000 (about a ninth of its present-day size) when the standard gauge railway arrived from Birmingham in 1840, four years after the Gloucester City Council had first met.

The line from the south was brought by the Bristol & Gloucester, formed in 1839 with power to absorb the Bristol & Gloucester Railway, incorporated in 1828, and to convert its ten miles of tramway from the Avon near Temple Meads to collieries at Shortwood, Parkfield and Mangotsfield. The tramway had carried coal which gave tremendous impetus to the early industrial growth of Bristol.

From Bristol's northern outskirts the line was to continue 22½ miles through flat farming country to Standish, where it was to join the broad gauge Cheltenham & Great Western Union, built by a nominally independent company across the Cotswolds from Swindon. The seven miles north to Gloucester were to be mixed gauge.

The Bristol company also got powers to use the C & GWU between Gloucester and Cheltenham, but that company's lease to Paddington from July 1843 placed the Bristol company in an awkward position, meeting the GWR broad gauge at both ends. It meant that if it adopted the standard gauge, a break of gauge would be necessary at Bristol and the GWR could thwart that by routing its traffic to and from the Midlands via Swindon. Yet if it remained friendly with the GWR, it could use its stations at Bristol, Gloucester and Cheltenham and avoid heavy capital expenditure. So the Bristol company subscribed to both the South Devon and Cornwall Railways.

The gauge question was unresolved when construction of the Bristol & Gloucester began in 1841 but earthworks were made to standard gauge dimensions. When the broad gauge was adopted, tunnels and bridges were never widened to provide the clearances of the London & Bristol.

The Bristol & Gloucester opened broad gauge on 8 July 1844 (the same day as the C & GWU), and got access to Temple Meads via a short spur from Lawrence Hill to the L & B just east of the station.

Fresh initiative by the GWR to drive the broad gauge to Birmingham (and the Mersey) was made in talks with both Gloucester companies, but when they reached financial deadlock, John Ellis, deputy chairman of the newly-formed Midland Railway stepped in following a chance meeting with two Birmingham & Gloucester directors on a London train. On his own initiative, he slightly improved the GWR offer, and by an agreement signed on 8 February 1845 the Midland leased both companies and absorbed them from 3 August. Besides removing the Midland's fear of broad gauge competition, the agreement dashed GWR hopes of having a broad gauge line to the West Midlands additional to the Birmingham & Oxford.

The third rail lingered on the Bristol & Gloucester in places, partly because the Avon & Gloucestershire Railway (a tramway

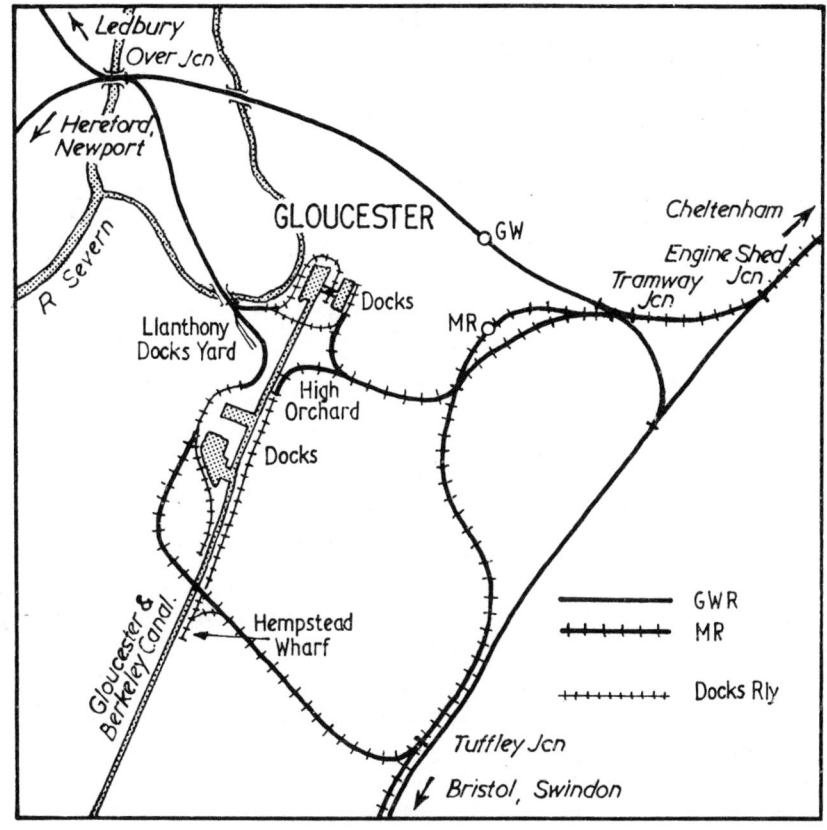

Ledbury
Over Jcn
Hereford,
Newport
R. Severn
GLOUCESTER
GW
Docks
Cheltenham
Engine Shed
Tramway Jcn
Jcn
Llanthony
Docks Yard
MR
High
Orchard
Docks
Gloucester &
Berkeley Canal
Hempstead
Wharf

	GWR
	MR
	Docks Rly

Tuffley Jcn
Bristol, Swindon

authorised with the Bristol & Gloucestershire in 1828 to get coal to
the Avon, in this case at Bitton), had running powers between
Mangotsfield and Coalpit Heath.

Gloucester's reputation for not having the best sited stations began
in 1844 when B & G trains had to use a platform attached to the
Birmingham company's station.

The GWR had problems, too, and from 1847—51 when Central
station ceased to be a terminus on the opening of the line to South
Wales, its famous 'T' station was in use on the Avoiding Line,
Gloucester coaches using a turntable on the main line to head
towards Central, where there were broad and mixed gauge
approaches. Even the Midland's original station meant through
trains reversing until it was replaced in 1896 by one with sharply-

curved platforms, which allowed North-to-South through running.
'A marvellous contrast to the old dustbin,' stated a travel guide.

The Midland was able to introduce its own standard gauge trains
to Bristol from 1854 through the opening of the TUFFLEY LOOP.
authorised as the Gloucester & Stonehouse Junction Railway in
1848 from Standish to Midland metals at Tramway Junction by
Gloucester Central. It laid its standard metals beside the C & GWU
between Standish Junction and Tuffley, where it took its own five
level-crossing route through Gloucester itself. The Loop cost
£159,042 and included a station at Haresfield, not served by the
GWR. It was on the side-by-side stretch on which, it was noted in
1902, 'the days of racing are long past.'

The Loop allowed the building of Eastgate station. Both closed
when Gloucester Central was modernised at a cost of £1¼ million
and officially opened in December 1975. Plans for an entirely new
station on both the Bristol and South Wales lines at Barnwood were
dropped because it would have been nearly a mile from the City
centre. But doubts linger because Gloucester now sees many
expresses taking the avoiding line. Reversal at Central, now simply
called Gloucester, is time-wasting—like break of gauge.

Brockthorpe, near Haresfield, was to have been the site of a £3
million marshalling yard, planned when a second Severn Tunnel
was under consideration, for the concentration of freight between
the Midlands, South Wales and the West of England. It was abandoned in 1964, the year after the Beeching Report.

Through the years the Bristol–Birmingham main line received
traffic boosts from the opening of the Mangotsfield–Bath branch in
1869, the Somerset & Dorset in 1874, and spurs at Westerleigh and
Berkeley in 1908. They were developed as a compromise to Midland
objection to a GWR proposal in its 1896 South Wales & Bristol
Direct Railway Bill for a branch from Chipping Sodbury to the
Severn & Wye Joint Line at Berkeley. It would have run parallel to
the B & G. Instead the GWR obtained running powers over the
eleven miles between the spurs.

BRISTOL: THE MIDLAND PRESENCE

The suburban approach was unbeloved of guide writers, *Baddeley* of
1902 noting: 'There is nothing of interest between Mangotsfield and

Bristol. Brickworks and other unsightly objects abound, and the last mile or two is through a network of rails and railway sheds.' They included Barrow Road locomotive depot (closed 1966), carriage sheds and repair shops and the one-platformed St Philip's Station, which *Bradshaw* noted as being about a mile by road from Temple Meads (although rather less if a crow flew across the Avon). It closed when Bath trains via Mangotsfield were diverted into Temple Meads in September 1953. A few sidings remain to serve Avon Wharf, reached by single line from Easton Road Junction (opened 1970) on the South Wales main line at Lawrence Hill. The rest of the area has become the Kingsland Industrial Estate.

Bristol–Gloucester stopping trains were withdrawn in January 1965, those to Bath (Green Park) surviving until the Somerset & Dorset closed in March 1966. The switching of Bristol–Birmingham expresses via Yate and Filton began on 5 May 1969 and Bristol–Mangotsfield closed completely at the end of the year, the Up line to Yate being retained for crew training. After a survey showed growing suburbs could not support a re-opened railway, BR offered the Bristol–Westerleigh–Bath trackbeds (sixteen miles) to five councils.

BIRMINGHAM & GLOUCESTER

The Bristol–Birmingham route gave the West Midlands a third trade outlet to a developing port, additional to those of Liverpool and London, to which it had been linked since 1838.

The Birmingham & Gloucester promoters went to Parliament in the same session as the Cheltenham & Great Western Union. The 45-mile line to Gloucester, authorised in 1836 from the London & Birmingham at Aston, was the most direct, missing small towns like Bromsgrove, Droitwich and Tewkesbury and virtually Cheltenham, which was to have only a branch, until local pressure forced the line to be taken closer. The unsatisfactorily-sited Lansdown station was the outcome.

With the Lickey Incline, the B & G was truly a line that went over the hills rather than round or through them. Most significantly of all, it ignored Worcester, an ancient city where iron-founding and other heavy industry had been established more than two decades before. The LNWR *Tourists' Picturesque Guide* of 1875, noted that glove-making gave employment 'to upwards of 1,500 hands, who send into the markets annually more than half a million pairs of gloves, chiefly kid and leather.' By that year the City had grown to 'a population of 38,116 souls.'

A legend that Worcester rejected overtures from the Birmingham & Gloucester must be debunked. Worcester people were pressing for closer ties with the West Midlands (nearer than Gloucester or Bristol) and were supporting a scheme, which proved to be ill-fated, for the Grand Connection Railway, from Abbot's Wood to Wolverhampton. Worcester, in fact, opposed a B & G Bill later because it did not include a branch to the City.

The Midland was quickly able to serve Worcester by taking advantage of the Oxford Worcester & Wolverhampton (page 87), which arrived from Abbot's Wood in 1850 and created a north–south loop with a 9½-mile line northwards via Droitwich (a Spa and a canal-connected port which could get its rock salt away in vessels of up to 600 tons) to Stoke Works, a now-demolished long-owned ICI salt works. The loop opened in 1852 and the Midland used running powers to serve Worcester. Three years later it closed a small station at Spetchley on the direct line, which had been its coach-head for Worcester.

The easier stretch of the B & G was built first: Cheltenham–Bromsgrove, thirty-one miles with 'first class' stations at both ends and several wayside ones. Gloucester and Cheltenham were linked in November 1840. While a locomotive and carriage shed was built at Cheltenham, the B & G established its main shed at Bromsgrove. This became a locomotive works, which survived until 1964 when it closed with the loss of more than 200 jobs. North of the Lickey Incline, the construction progressed through expanding outer Birmingham (Vol 7) until link-up with the London & Birmingham at Curzon Street station in August 1841.

Ever since, the Lickey Incline has been a magnet for enthusiasts. A 1945 memory: outwards from Birmingham on the 12.32pm stopping train to Gloucester with a Class 2 4-4-0 and four compartment coaches. Back on the 12.47pm ex-Hereford—banked by 0-10-0 'Big Bertha'. Unique . . . glorious bark . . .

Yet while providing delight for generations of enthusiasts, the incline, at 1 in 37 the steepest on a British main line, posed terrible problems for the operators. In 1933, the LMS quadrupled the 2¾ miles between Bromsgrove and Stoke Works to ease congestion at

MIDLAND

QUICKEST AND MOST DIRECT ROUTE
BETWEEN
BIRMINGHAM
AND
Cheltenham, Gloucester, Bath, Bristol,
AND
WEST OF ENGLAND.

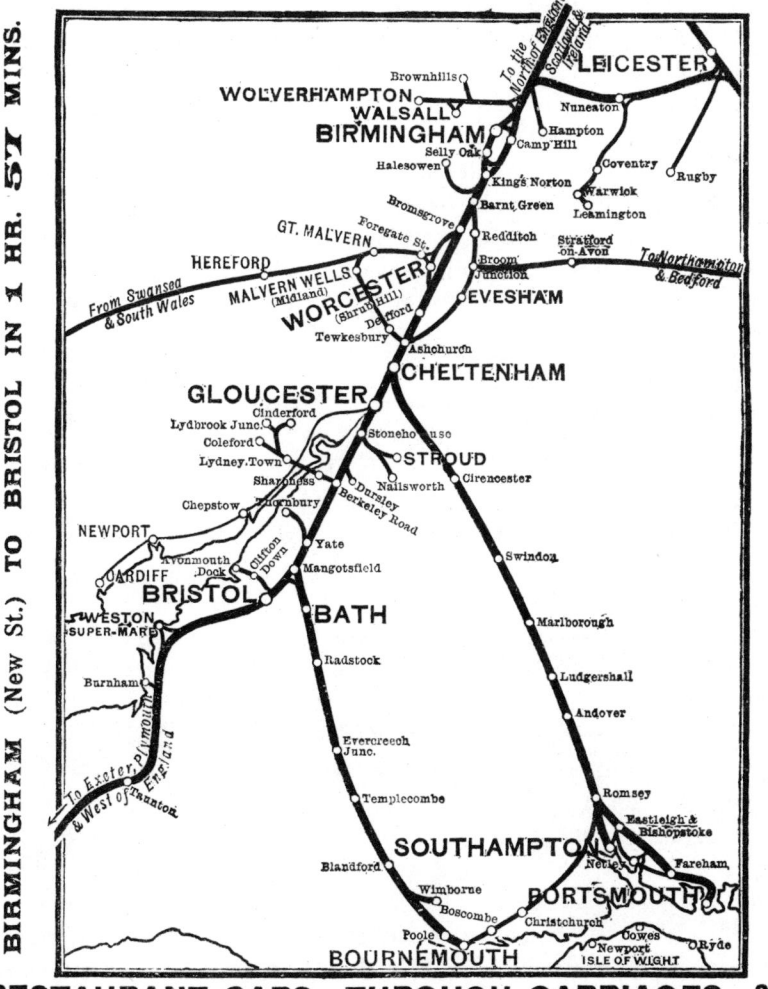

RESTAURANT CARS, THROUGH CARRIAGES, &c

W.B. 67/1914. (3581).

the foot of the bank and installed colour-light signals. Nowadays the incline is a symbol of diesel superiority over steam, for the speed limit for ascending trains has been more than doubled from 30mph to 75mph, and most expresses climb without bankers which, however, assist about two-thirds of freight trains.

To meet wartime demand, the Bristol–Birmingham main line was quadrupled between Gloucester and Cheltenham from 24 August 1942, the scheme including an extra down goods loop between Engine Shed Junction (as Gloucester Sidings signal box was re-named on completion of the work) and Tramway Junction. The main line reverted to double track in 1967. That was two years after the BRB had examined freight trends in a report on *The Development of the Major Railway Trunk Routes*. Three lines between the West Midlands and South Wales and the South West were analysed. Birmingham–Bristol via Bromsgrove, the shortest: via Worcester and Hereford, a route of about the same distance between Birmingham and Newport, but handicapped by not offering a direct link to Bristol; and Stratford-upon-Avon–Cheltenham, which was devoid of major centres of industry.

In 1964 the routes were carrying just over 250,000 tons of freight a week: that via Lickey, 90,000 tons; the Worcester and Hereford, 70,000 tons, and the Honeybourne link, 60,000 tons.

The planners favoured that via Lickey with its tailor-made connections at Birmingham with the main line into Yorkshire via Derby; at Gloucester for South Wales, and at Bristol for the route between London and the West, which the BRB had selected for development.

The Report labelled Newport–Shrewsbury–Crewe as a duplicate, but it was one on which more freight was concentrated after the closure of secondary lines, including the Mid Wales and the old Midland route between Swansea and Hereford.

Bristol–Birmingham line capacity was increased after the withdrawal of local passenger trains between Bristol–Gloucester–Birmingham in the same year. The next generation of Inter-City expresses will be HSTs. More locally, most trains on the Birmingham Cross-City line, successfully developed in the 1970s, turn round at Longbridge, but several were experimentally extended to Bromsgrove in 1979, and a bus link introduced between station and town centre.

Plate 9

Above: Ever anxious for publicity, the Great Western Railway in the 1930s claimed the title of the world's fastest train for the up afternoon Cheltenham Spa Express which became known as the Cheltenham Flyer when it was timed to run the 77 miles from Swindon to Paddington in one hour. The train is seen here near Haresfield headed by the unique semi-streamlined Castle class 4-6-0 No 5005 *Manorbier Castle* on 30 April 1936. (*L&GRP, courtesy David & Charles*)

Plate 10

Below: Heavy Cotswold gradients often meant double-headed expresses. Nos 2945 and 7006 on a Cheltenham-Paddington train in August 1950 pause at Kemble. This was the railhead for long distance commuters and junction for Tetbury and Cirencester Town (branch platform right) and had an excellent rail service out of all proportion to its size as a village. (*Real Photographs*)

Stations at Cheltenham
Plate 11
Above: A general view of Cheltenham Spa St James taken in April 1932. This was the Great Western terminus for trains to London. (*L&GRP, courtesy David & Charles*)

Plate 12
Below: Cheltenham Spa Malvern Road station looking south west towards Lansdown Junction. Great Western trains from St James also used this station but its most important function was to serve trains running to Birmingham via Stratford-on-Avon. (*L&GRP, courtesy David & Charles*)

BATH AND OTHER LINES

With some justification, the Midland always regarded Mangotsfield–BATH as a main line rather than a branch, mainly because it was the gateway to the Somerset & Dorset. For years, railways boosted Bath's tourist trade, the LNWR and Midland running day excursions to Bath Races from Coventry and other parts of the Midlands —they called at Kelston as well as Bath.

Once it had gained control of the Bristol main line, Bath was an obvious target for the Midland, and George Hudson presided when the board decided in 1846 to promote the route as one of its earliest ventures. The economic climate killed the Bill, and Bath was growing fast by the time the branch was authorised in 1864. A decade later Bath's population of 54,000 was 3,000 more than Cheltenham's and 18,000 more than that of Gloucester. The Bath branch opened to passengers on 4 August 1869 and to goods on 1 September.

A temporary terminus at Queen Square, Bath, was replaced by the 'Midland Station' a year later; it was not called Bath Green Park until 1951. An imposing frontage gave it a deceptively large appearance and misled one reputable guide book to describe it as 'one of the largest stations in the West of England,' despite having only two platforms and two middle roads under its arched roof.

No one welcomed the Midland to Bath more than the Somerset & Dorset directors, urgently seeking an outlet north because the GWR blocked its projected approach to Bristol. A 25-mile through-the-Mendips branch opened in 1874 to a junction half-a-mile west of Bath Midland, and the company moved its offices to Bath from Glastonbury three years later.

The Midland's Bath–Mangotsfield–Bristol (St Philip's) route was three miles longer than the GWR's. The Bath branch was upgraded to the main line standard in the 1930s because of its strategic importance as a north–south route. Until then, the heaviest locomotives allowed were Class 3 4-4-0s. Strengthened track and bridges allowed S & D 2-8-0s to turn on the Mangotsfield triangle.

Although the branch kept passenger services until 1966 and freight until 1971, Kelston station (without road access, yet popular with anglers and racegoers) closed in 1949. Bitton station and yard is now the headquarters of enthusiasts who formed the Bristol Suburban Railway.

Authorised with the Bath branch in 1864, the 7½-mile Yate–
THORNBURY branch opened 2 September 1872, closed to
passengers in 1944 and completely in 1966. Six miles to Tytherington
Quarry were relaid with track from the Bath branch and since 1972
block trains carrying 1,000 tons have been running from the quarry,
among the biggest in the West, from which stone for the King
Edward Dock, Avonmouth, had been blasted nearly seventy years
earlier.

The 2½ mile Coaley–DURSLEY branch was the work of local
businessmen in a valley famous for its woollen and carpet mills.
They formed the Dursley & Midland Junction Railway in 1855 and
faced with 'no peculiar engineering difficulties or tunnels and with
gradients and curves generally favourable' (*Bradshaw's Manual* 1859),
accomplished construction in a year, opening to goods 25 August
1856 and passengers 18 September. The Midland worked the line,
absorbed the company in 1861, and the line continued to carry
passengers until 1962 and freight until 1970.

The Stroud Valley developed cloth and a host of other small
industries long before railways, its first improvement in transport
being the Stroudwater Canal of the late 1700s, which allowed small
boats from the Severn to reach the town. In the 1860s, the Midland
challenged the GWR monopoly at Stroud with a branch off the
STONEHOUSE & NAILSWORTH Railway, a 5¾-mile line
opened 4 February 1867 to Nailsworth, a small town with a variety
of industry, notably cloth, but including leather goods and rugs. The
1¼-mile offshoot from Dudbridge to Stroud Cheapside, opened
16 November 1885 by the Midland, which had bought the S & N in
1868, was no match for the GWR's Gloucester–Swindon main line.
'The Midland's service to Stroud is somewhat poor, and does not
compare at all favourably,' noted *The Railway & Travel Monthly* in
1914. 'The branch is conducted by the MR standard 0-4-4 side tanks
and types of 0-6-0 goods tender engines with outside cranks, there
being usually three engines in steam at a time. They are stationed at
Gloucester shed.'

Withdrawal of the Stroud and Nailsworth passenger trains in June
1947 was one of the last LMS economies. Initially stated to be
temporary, the economies were confirmed as permanent by BR two
years later. Both branches closed completely in 1966.

In 1980, Gloucester celebrated its 400th anniversary as a port, having been granted a Charter by Queen Elizabeth I. What came to be known as the 'most inland natural port of England' developed between 1827–92 and was well served by the GLOUCESTER & CHELTENHAM TRAMROAD of 1811, which played a major role in Cheltenham's development through the transport of raw materials, including roadstone and coal. Gloucester docks became overshadowed by those of Sharpness, which developed to take 9,000-ton ships, while Gloucester could only handle those of up to 1,200-tons. Railways were important to both ports. The Midland served Gloucester docks from 1848 and after the GWR's arrival from Over Junction in 1854, handled traffic from east and west docks under agreements often disputed.

The Midland had two lines stemming from the Tuffley Loop: the HIGH ORCHARD line, which served the Gloucester Carriage & Wagon Company, established near the docks in 1860, and the 1½-mile HEMPSTEAD WHARF branch, which served the Gloucester Gas Light Company. Branches and the sidings and small yards which they served closed between 1938–71. Despite their comparatively small size, the docks handled large volumes of traffic, much of it bound from overseas to the West Midlands, the total passing the million ton mark before World War I.

Important small towns like Tewkesbury which had up to thirty daily stage coach services suffered when they were abruptly withdrawn as railways opened. But Tewkesbury was luckier than some towns, for the Birmingham & Gloucester had opened a 1¾-mile branch from its main line at Ashchurch on 21 July 1840, although the historian John Norris believes it is likely the branch was completed much earlier, being used to convey materials for construction of the main line, delivered to Tewkesbury by river craft. Eventually the branch extended from ASHCHURCH to GREAT MALVERN, the gap being filled by the Tewkesbury & Malvern Railway of 1860. It opened a single line from a new station at Tewkesbury to the GWR at Great Malvern on 16 May 1864. The T & MR, an independent company whose records have not survived, was absorbed by the Midland in 1877. Weekday passenger services were cut back from Great Malvern to Upton-on-Severn in 1952 and the stretch of almost seven miles closed completely. The remainder retained passenger trains until 1961, following strong local protest that Tewkesbury's tourist trade would suffer without them. Tewkesbury–

Upton closed completely in July 1963; Ashchurch–Tewkesbury in November 1964. At Ashchurch the Malvern branch was linked by a spur crossing the main line on the level with the BIRMINGHAM & GLOUCESTER LOOP, the outcome of pressure from in-dustrialists at Redditch and, further south, farmers in one of Britain's richest agricultural areas. Both wanted quick transport to market and found the main line too far away to be satisfactory. Redditch has steadily expanded as a New Town since 1964 and plans to have a population of some 70,000 soon. Yet the RED-DITCH RAILWAY, opened from Barnt Green in 1859 (passengers 19 September; goods 1 October), now has only a single platform for diesel multiple-units on the Birmingham Cross-City link. Long gone is the goods yard, which handled dozens of local industrial products including motor cycles, batteries and fishing tackle. In its place a roadstone terminal is in use.

The EVESHAM & REDDITCH Railway was authorised in 1863 to close the 17¼-mile gap in the Loop between the towns, the Mid-land having linked Ashchurch and Evesham (ten miles) on 1 October 1864. The E & RR was completed in two stages. Evesham–Alcester was opened to goods on 16 June 1866 (passengers 17 September), and Alcester–Redditch on 4 May 1868. The Loop, never engineered to express standards, was used by goods trains to avoid the Lickey Incline, and achieved importance during World War II when the Stratford-upon-Avon & Midland Junction was improved to keep freight trains clear of the West Midlands, a south-facing exit spur being put in at Broom. For years the only connection between the OW & W and the Loop at Evesham was through sidings, but a running junction was provided soon after Nationalisa-tion for Honeybourne–Cheltenham route diversions. Ashchurch–Redditch passenger services were withdrawn in 1963, eight months after buses were substituted for trains between Evesham and Redditch, and freight services were withdrawn 1963–64.

THE TRUE PRICE OF GWR FAILURE

The price that the GWR eventually paid through the Midland snatching control of the Bristol–Birmingham route in 1845 became obvious about half a century later when it embarked on a rival route, which suffered from being ten miles longer and running through mainly rural areas where intermediate passenger traffic potential was limited. The route of 1908 was forged from lines old and new, remarkable for their variety. The GWR used its own lines from Bristol to the outskirts at Westerleigh where a spur had been opened to the Midland main line near Yate at the same time as the Badminton cut-off on 1 May 1903. The GWR Birmingham trains used the Gloucester Avoiding Line (which had been re-opened to goods on 25 November 1901), the Cheltenham–Honeybourne route (new), the Oxford Worcester & Wolverhampton's Honeybourne–Stratford-on-Avon line (old), the Stratford-on-Avon Railway (old) to Bearley, and the Birmingham & North Warwickshire (new).

The Cheltenham–Honeybourne line was something of a protective venture by the GWR, forced upon it as companies attempted to reach Birmingham from the south. The Birmingham, North Warwickshire & Stratford-upon-Avon Railway of 1894 was to enable the Manchester Sheffield & Lincolnshire to reach Birmingham using the East & West Junction between Woodford and Stratford. The plan was abandoned when construction powers of the Warwickshire company passed to the GWR in 1900.

Two years earlier, the Andoversford & Stratford-upon-Avon Railway was proposed with the object of exploiting the Warwickshire line. It was to be worked by the Midland & South Western Junction. It was rejected by Parliament only after the GWR agreed to develop Cheltenham–Honeybourne and double between Stratford and Bearley.

Completed on the seventh anniversary of authorisation on 1 August 1899, the 21¾ miles between Cheltenham Spa (Malvern Road East Junction) and Honeybourne (East and North Junctions), had been progressively built under Cotswold Edge in short sections. The Honeybourne line had its own station at Cheltenham—Malvern Road—opened 30 March 1908 so that through expresses would not have to reverse at St James'. The line was liberally provided with stations and halts, including Racecourse stations at Cheltenham (just over two miles north of Malvern Road) and Stratford (just south of the town).

The Midland fought hard to make the GWR development as difficult as possible, notably through a legal battle when the service was ready for opening in 1908. It maintained that the GWR trains

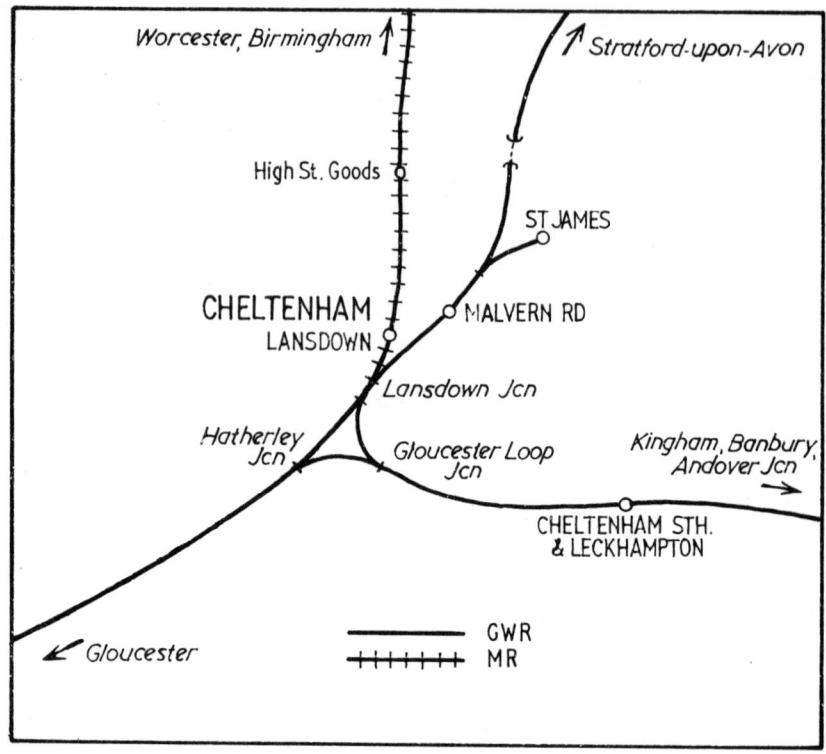

should not use the Westerleigh spur, but run into Bristol via Mangotsfield, filling the Midland coffers. The GWR won an Appeal Court ruling and in November 1908 switched expresses which it had introduced the previous July. The next summer three expresses established a pattern between towns beyond Wolverhampton and Bristol. The Shakespeare Express of 1910 ran between Birkenhead (Woodside) and Plymouth (Millbay). The Honeybourne Act authorised doubling between Honeybourne and Bearley.

HONEYBOURNE—STRATFORD-UPON-AVON

The first nine miles were those of the OW & W branch, authorised in 1846, but not opened until 12 July 1859. It was the last of four branches the Company completed before becoming the West Midland Railway. Under that guise the Company completed a short connection with the Stratford-on-Avon Railway, incorporated 10 August 1857 between Stratford and the Birmingham & Oxford at Hatton, nine miles. It opened as a single mixed-gauge line on 10 October 1860, the half-mile link at Stratford following on 24 July 1861.

One of the intermediate stations that I got to know, and used during a family holiday was Wilmcote, which serves the village where Mary Arden, Shakespeare's mother was born in a cottage just down the road. In late August 1945, I sat on the platform and watched heavy freights being banked from Stratford, at their heads the distinctive Aberdare Class 2-6-os, with connecting rods thrashing round outside the frames. A soldier got off a Leamington Spa–Stratford GWR railcar and I wondered if he was on leave, or finally coming home from the war to this, the glorious heart of England, into which the GWR seemed to fit so perfectly.

World War II gave a boost to this mainly freight artery, primarily because it helped to keep traffic clear of congested Birmingham; in 1939 doubling took place between Bearley and Hatton, Bearley–Stratford having had double track since 1907. A development at Bearley associated with the Bristol–Birmingham route involved a re-sited junction for the $6\frac{1}{2}$-mile ALCESTER RAILWAY, necessary when the Birmingham & North Warwickshire line (Vol 7) was constructed 1907–08.

The Alcester Railway, authorised in 1872 and opened 4 September 1876, had a chequered career due to the world wars. The branch, which ran to Alcester on the Birmingham & Gloucester Loop, closed during World War I, re-opened 1922–23, and lost regular passenger services in September 1939, but carried workmen's trains 1941–44 after the Maudslay Motor factory moved to Great Alne following the Coventry blitz. The branch was a wagon store for some years after complete closure in 1951.

BRISTOL–STRATFORD–BIRMINGHAM: LATER YEARS

Several daily expresses once linked Birmingham (Snow Hill)–South Wales, Bristol and the West via Stratford, perhaps the most intersting being the innovatory GWR diesel railcar Birmingham–Cardiff expresses introduced 1934. A quarter of a century later, BR's

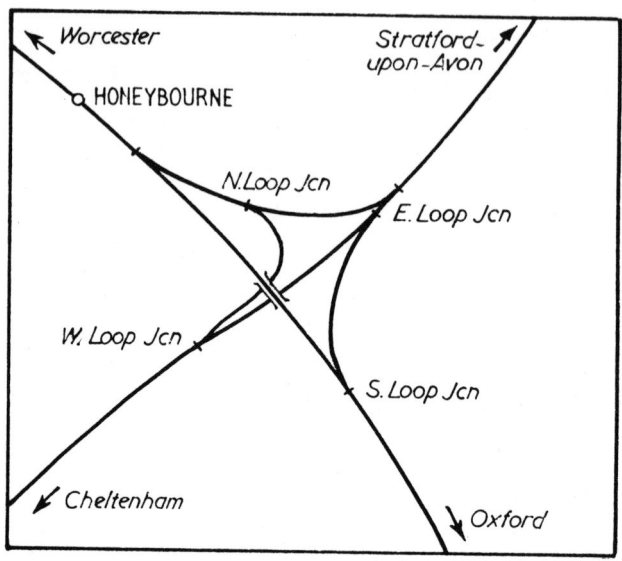

attempts to meet motorway competition with high-speed Bristol–
Birmingham services via Lickey, and declining local traffic via
Stratford, led to route decline. Cheltenham (St James')–Honeybourne local passenger services ceased in March 1960, although
Honeybourne–Stratford proved sturdier, surviving until May 1969,
when Honeybourne station closed as well, the East Loop following
in November 1970. The route was retained for a reduced number of
through services until 1968, the last being Gloucester–Leamington
Spa, maintained appropriately by a single diesel railcar. In 1973,
only enthusiast support prevented a sponsor from losing heavily on a
trial summer weekend Gloucester–Stratford service. Heavy rain
proved the unpredictability of tourist weather—and the infallibility
of BR's economy case.

Ironically, severing the Bristol–Birmingham alternative route
took place at a time when the railways had plenty of spare traffic
capacity, the roads virtually none, and millions of pounds were
about to be spent widening the congested M5 motorway.

Had rail freight been booming the line might have survived,
because in June 1960 Racecourse Junction was opened to provide a
west-to-south outlet for the Stratford-upon-Avon & Midland
Junction route (another West Midland avoider) which ran further

NEW THROUGH EXPRESS SERVICE
BETWEEN
BIRKENHEAD, CHESTER, SHREWSBURY, WOLVERHAMPTON, BIRMINGHAM, STRATFORD-ON-AVON, CHELTENHAM & BRISTOL,
AND THE
WEST OF ENGLAND.

CORRIDOR CARRIAGES,

1st & 3rd Class.

✗ ✗ ✗ ✗ ✗ ✗ ✗ ✗

WITHOUT CHANGE OF CARRIAGE.

✗ ✗ ✗ ✗ ✗ ✗ ✗ ✗

Luncheon and Tea Cars.

To and From the Cornish Riviera.

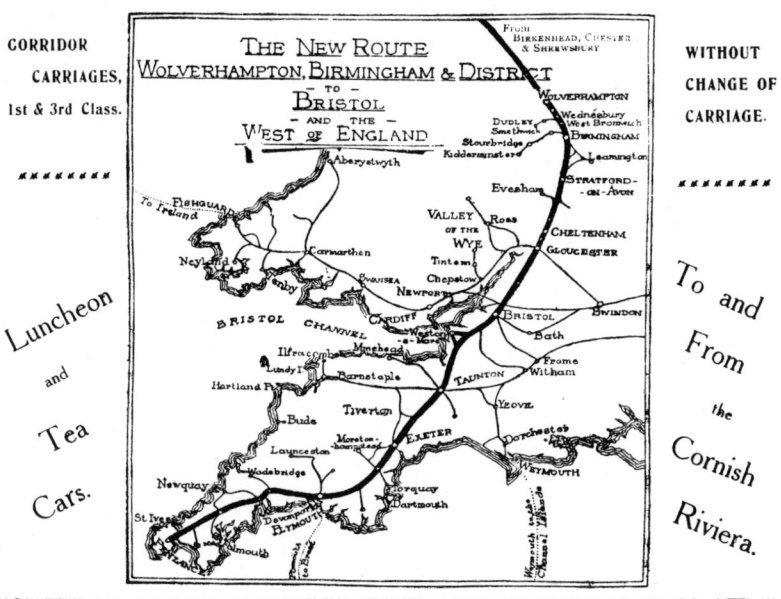

THE NEW ROUTE WOLVERHAMPTON, BIRMINGHAM & DISTRICT — TO — BRISTOL — AND THE — WEST OF ENGLAND.

From BIRKENHEAD, CHESTER & SHREWSBURY

WEEK DAYS.

E—North Road.   P—Depart 3 40 p.m. on Saturdays.   ‡—Give notice to Guard at Shrewsbury.

west to join the Birmingham & Gloucester Loop. Racecourse Junction was opened for just five years, closing after Cheltenham–Honeybourne was eliminated by BR from freight development in favour of Lickey. But the section lingered on until blocked by a freight train derailment at Winchcombe in 1976. Later the track was lifted and the route became one of the few notable closures of recent years.

Today, Stratford is a southern outpost of the West Midland commuter service, not as well served as stations on the North Warwickshire line closer to Birmingham. Although Stratford is England's second most important tourist centre, BR finds difficulty in attracting sufficient tourist traffic to serve it well: an excursion I used in 1974 was poorly supported, possibly because it took almost $3\frac{1}{2}$ hours from Manchester Piccadilly via Dore & Totley. For London orientated visitors to Britain, BR uses Coventry as a railhead for Shakespeare Country coach tours.

CHAPTER V

Main Lines to South Wales

CHELTENHAM & GREAT WESTERN UNION

My RAF uniform was magic. The first time I crossed the Cotswolds from Reading to Gloucester to join my parents on holiday at Ross-on-Wye, the express was full, but I was invited into a reserved compartment (first class, I think) and given a window seat. More than thirty years later I remember that hospitality with gratitude in the same way as I still regard Ross-on-Wye as a town of happy memory.

The C & GWU had a reputation of being a friendly line, as Norman Shrapnel pointed out when he wrote in *The Guardian* (21 June 1976) on how the Cotswolds, having been urbanised, were then being suburbanised:

> . . . And although Kemble Junction is not what it was, they still manage to preserve a decent sense of priorities.
>
> When Kemble really was a junction it also served as a symbol of social harmony, a lesson to us all in the art of living together. From this railhead the rural classes contentedly went their various ways—the rich commuters away on the main lines to their chambers and clubs, with their tweedy wives and daughters off for a little basic marketing at Fortnum's, while the poachers, pig-food men, village mums, and the rest of the peasantry shuttled around the convenient local halts in the vicinity of Cirencester and Stroud.

Norman Shrapnel remembered a whim of the wealthy Cotswold commuters:

And what became of those fastidious breakfasters who used to bring their own new-laid eggs, to be cooked on the train to their strict specifications? If they have passed on, God will no doubt have had a sharp word for people who couldn't even trust the old Great Western.

Although overshadowed by the London & Bristol, the C & GWU is very much vintage Great Western, Swindon–Gloucester being among three 'probable' branches mentioned in the original 1833 Prospectus. It was envisaged by the GWR directors as a line far more important than the aim for which it was promoted in 1836: to get Cheltenham's rich to London quickly. For all but six of its forty-two miles from the London & Bristol at Swindon to Cheltenham were forged into the first main line between London and South Wales, while the rest (Gloucester–Cheltenham) became part of the original West Midlands–South Wales route. The Cotswold line remained the only London–South Wales route until the Severn Tunnel of 1886, when it was relegated to what it has been ever since—a valuable subsidiary route.

Unlike the Bristol main line, the Cheltenham route was born amid strong opposition from the London & Birmingham, which had territorial ambitions to strike west from Tring via Aylesbury, Thame, Witney and Burford to Cheltenham, plus an Oxford branch. It would have put Cheltenham under 100 miles from Euston (99 miles actually) compared with 120 to Paddington, although gradients would have been stiffer. That scheme died and opposition of the Thames & Severn Canal Company overcome, while agreement was reached with the Birmingham & Gloucester about ownership of the Cheltenham–Gloucester route, for which both companies went to Parliament at the same time. They agreed to the joint purchase of the 9-mile Gloucester & Cheltenham Railway, so-called by its Act of 1809, but as D. E. Bick explains in a detailed book with that title, it was usually known as the Gloucester & Cheltenham Tramroad. And so it was: a horse-drawn plate tramroad, which from its instigation by the ever-questing Cheltenham landowners, wanting to get coal and goods from Gloucester docks, played an important part in the development of the Spa.

The Birmingham promoters took over the section from Cheltenham from the C & GWU and reached Gloucester from the north in 1840, ahead of their own completion into central Birmingham.

Between the towns, ownership was divided, the Midland, as in-
heritors of the B & G, taking charge between Gloucester and a point
west of Churchdown, the only intermediate station. That became
GWR property, with the line to Cheltenham, although the station
was jointly staffed.

The thirty-six Cotswold miles between Swindon and Gloucester
can be considered in two halves. The first was formed by the eighteen
miles between Swindon and Cirencester, completed 31 May 1841
(on the same day as the Bristol main line reached Chippenham from
Paddington). Kemble did not become a junction (nor Cirencester a
branch) until the C & GWU reached the B & G at Standish Junction
four years later.

By then the Cheltenham company had been in GWR ownership
for two years. Dropping down from the Cotswolds by the Stroud
(Golden) Valley to the Severn Valley involved heavy engineering
works including Sapperton Tunnel (1,864yd) and nine viaducts,
constructed originally from plentiful local timber. These works
accounted for the sixteen miles taking four years. Mid-way through
Sapperton Tunnel a three-mile descent to Brimscombe began on
gradients of 1 in 75/60; gradients which still require most freight
trains to be diesel-banked, making the route a none-too-popular
alternative to the Severn Tunnel.

The C & GWU terminus was at Cheltenham Spa St James'
(sometimes with an apostrophe, sometimes without), built after the
1844 Amalgamation Act with the GWR provided for a 1¼-mile
extension from Lansdowne Road station. It put St James' station
nearer the town centre than any other.

The Cotswold route had its first major change when it was converted from broad gauge to standard in 1872 and it suffered its first
traffic loss in 1890 when London–South Wales expresses were finally
diverted through the Severn Tunnel (opened four years earlier) and
by reducing journey times by thirty minutes, put the Welsh Capital,
Cardiff, within four hours of England's.

St James' was the terminus of famous and interesting trains,
notably the Cheltenham Flyer, which reflected the importance
which the GWR attached to the spa. Introduced only months after
Grouping (9 July 1923), it enabled Britain to claim possession of the
world's fastest regular steam train, though all the speeding was done
away from the Cotswolds, the 77¼ miles between Swindon and

Paddington being booked in 65 minutes at an average start-to-stop speed of 71.3mph. In more recent years there was the Cheltenham Spa Express, a title withdrawn when the train was decelerated. Today, Cheltenham can be reached from Paddington in about 2 hours 20 minutes with stops at Swindon, Kemble, Stroud and Gloucester, about the same as the best timings before World War II. St James' more humble long-distance services were those over the Midland & South Western Junction (page 163). It was a busy station almost up to the end and in June 1958, only three years before it lost its Southampton trains and Kingham local service, it issued more than half a million tickets.

The Cheltenham Flyer may have been the most famous steam train to run through the Stroud Valley, but the best remembered ones will be the railmotor services tailor-made for the Valley, packed with industry where workers travelled short distances. The railways were of great value to most local industries—cloth, leather-board, carpet and flock mills, iron and brass foundries, hairpin, umbrella and walking stick factories; breweries, quarries, brick-works, dye and paint works, timber yards.

Of the four-mile Stroud–Chalford service, *The Railway & Travel Monthly* noted in January 1914 that

> It would be difficult for anyone not conversant with the locality to appreciate the boon to the district that the introduction of a railmotor service has proved. Previously, the only means of transport between Stroud, Brimscombe and Chalford for the workers in the mills and factories in the valley was an antiquated horse 'bus, the stations being too far and the stopping trains too infrequent to tap this valuable source of traffic.

Yet if trains replaced horse buses, this was an area where the GWR experimented with its own buses and before World War I it was running a service from Stroud to Cheltenham and, for tourists, to the pretty village of Painswick.

Buses (and cars) were destined to displace the Gloucester–Chalford auto-trains from 2 November 1964, the date on which the three intermediate stations between Swindon and Kemble were closed. At Cheltenham, St James' and Malvern Road stations and the short line between them, closed on 3 January 1966. The sixteen miles between Swindon and Kemble were singled in 1968.

Stations between Kemble and Gloucester were noted for handsome stone buildings and a major conservation row developed when BR announced plans to close Stonehouse from 6 October 1975. Brunel's station had by then become a building listed as being of architectural or historical interest. The Parish Council agreed to pay £13,000 towards modernisation and maintenance of local services, which were subsequently improved, and in 1978 the C & GWU was given the first refurbished diesel multiple-units to go into service in the West of England Division. Early in 1981 the line was further down graded by the withdrawal of some Paddington–Cheltenham expresses, forcing some passengers to change at Swindon. During the economies of the 1960s, the change on the line was reflected by a pub at Purton, which used to be the first station out of Swindon. It was re-named from 'The Railway' to 'The Ghost Train'.

SOUTH WALES RAILWAY: GLOUCESTER–NEWPORT

Every day sees an increased supply of coal thrown into the London market, and every year sees fresh collieries opened to meet the demand. The total number of pits in South Wales is about 330, which produce between 7 and 8 million tons annually.

Murray's *Hand-book for South Wales* of 1860 was reflecting the booming coal trade only eight years after completion of the 'great trunk line from Gloucester to Milford Haven.' Although the Midland halted the GWR broad gauge advance from Gloucester to Birmingham, Paddington made no mistake when it came to carrying the broad gauge into the heart of South Wales and beyond to exploit Irish traffic. By the time the C & GWU reached Gloucester in 1845, construction had already started. The first intention of the South Wales Railway was to head directly towards South Wales from Standish Junction, avoiding Gloucester by bridging the Severn Estuary near Awre.

But Gloucester business interests were too powerful to allow that, and sought allies among industrialists in Monmouthshire who wanted better rail facilities by taking the line up the Usk Valley from Newport, and to Gloucester via Monmouth, adding eighteen route miles and numerous stiff gradients. The Admiralty stepped-in, refusing to allow the Severn to be bridged for fear of causing interference with coastal navigation. The South Wales Railway Act of 1845 for a line from Fishguard and Pembroke was to carry it no further east than Chepstow, so the question of an Estuary crossing was left unresolved.

There was to be a branch from Newport to the Monmouth & Hereford at Monmouth. It was never started, neither were a Severn bridge or tunnel, and when it became obvious that the route would be built along the river bank between Chepstow and Gloucester, success was assured by a GWR-supported company, the Gloucester & Dean Forest Railway. It was promoted, with headquarters at Ross, by Cheltenham residents concerned at the high cost of transporting coal from the Forest of Dean. Incorporated in July 1846 from Gloucester to the proposed Monmouth & Hereford at Grange Court Junction, about eight miles west, it was also projected for another $7\frac{1}{2}$ miles west to reach the SWR at Awre. Both companies got powers for this, but the SWR took control of construction west of Grange Court.

In 1847, the G & DFR got an Act for a $1\frac{1}{4}$-mile branch from Over Junction on the outskirts of Gloucester, to Llanthony (the name belongs to an ancient Priory), with powers to build a dock or basin. But after running out of money, it left construction to the GWR, which opened the branch on 20 March 1854, the month following its lease of the G & DFR. The branch, which after crossing the Severn spread out in sidings, remained broad gauge until 1869. Six years later the GWR absorbed the G & DFR.

CHEPSTOW: WYE BRIDGE

The first section of the SWR stretching seventy-five miles from Chepstow to Swansea opened in June 1850, but work on Brunel's Wye Bridge at Chepstow fell behind schedule and the $26\frac{1}{2}$-mile line from Gloucester, including the G & DFR did not open until 19 September 1851, and then only to a temporary terminus a mile east of Chepstow. The main line was linked when the bridge opened 19 July 1852, doubling taking place the following April. The suspension bridge, topped by two towers 150ft above low water level lacked the grace that Brunel achieved with the Tamar Bridge, which he designed with experience gained at Chepstow. While 'scarcely

Plate 13
Above: The Midland & South Western Junction never lacked enterprise in through services: the 1.10pm Cheltenham–Southampton West (the South Express) of 1914, seen at Marlborough, included a through coach from Manchester. (*Ken Nunn collection, courtesy Locomotive Club of Great Britain*)

Plate 14
Below: Andoversford Junction with the Midland & South Western Junction Railway branching to the right towards Swindon leaving the Great Western cross-country line to the left heading for Kingham and Kings Sutton. (*L&GRP, courtesy David & Charles*)

Plate 15
Above: Cross-country junction – Honeybourne, with its four platforms serving trains on both the Oxford and Worcester line and also coming off the Cheltenham–Stratford-on-Avon line. (*L&GRP, courtesy David & Charles*)

Plate 16
Below: Hereford Barrs Court station on the Shrewsbury & Hereford joint line, forming the cross-country main line from Newport to Shrewsbury and Crewe. (*L&GRP, courtesy David & Charles*)

harmonizing with the rest of the scene' (Murray's *Hand-book* again), the Wye Bridge remained in its original form until rebuilt in 1962, after years of 15mph restriction. The tubes and towers were removed and the train deck supported by a deep lattice truss.

Gloucester's fortunes as a railway centre were strongly influenced by the South Wales route. The Gloucester & Dean Forest company, based at Ross-on-Wye, exhausted its share capital on limited construction and left the GWR to build a branch to Gloucester Docks, now the only surviving one. It opened in 1854. Gloucester Central station's long, single platform, dates from 1852. Its sister, a shorter Up platform, was added in 1889. Since concentration of passenger trains on the original platform in 1975, the Up platform has been used only for parcels.

YEARS OF CHANGE

The station modernisation marked the end of almost fifty years of change which began when the Severn Tunnel marshalling yard was enlarged under a Government unemployment relief project in the early 1930s—the same one that provided impetus for extension of Temple Meads station. In summer 1941, the ten miles between Severn Tunnel Junction and Newport were quadrupled for wartime traffic and afterwards, Bishton Flyover was built to carry the up slow line across the main lines, so that both slow lines ran alongside the much-extended Llanwern Steelworks. In later years the picture has been mainly of decline, and destruction of the Severn railway bridge meant little, if any, extra traffic.

In winter 1964, Cheltenham–Gloucester–Swansea passenger services were severely pruned, thirteen stations and halts being closed. Four survive between Gloucester and Newport, namely Lydney, Chepstow, Caldicot Halt, and Severn Tunnel Junction. At Severn Tunnel Junction, a vicar protested when BR introduced the bi-lingual sign of Cyffordd Twnel Hafren in 1975. The Reverend Christopher Gwilliam said the place never existed before the GWR was built and it had never been Welsh-speaking. Caldicot Halt, just above Severn Tunnel mouth, was reprieved after a TUCC Inquiry, which found that closure could cause local road congestion. Gone, too, are branch services that used part of the South Wales main line:

Gloucester–Cinderford, Ledbury, Hereford (via Ross), and Chepstow–Monmouth. But the main line survives, carrying a South Wales–North East Inter-City service, and trains diverted when the Severn Tunnel is closed, and some London–South Wales freight services.

BRISTOL & SOUTH WALES UNION

Although the Bristol & South Wales Union was a company far grander in title than achievement, it did pioneer a route between Bristol and Severn shore which the GWR exploited to the full once the Tunnel was built. The B & SWU was born from one of many schemes to conquer the wide estuary with its fast-flowing tides and shifting sands, by either bridge, tunnel or ferry. Among the most ambitious (and impracticable) was the Bristol & Liverpool Junction Railway of 1845, which envisaged a bridge from Aust Cliff to Beachley, including four spans of 1,100ft. A year later, the first company known as the B & SWU was formed. It bought the existing ferry from New Passage to Portskewett on the north bank, and some land, but was soon dissolved.

Brunel became engineer when the project was revived in 1857, but by then he was a sick man and this prevented him doing much. The company quickly reached agreements with landowners for 'the land which will be required for the tunnel, where the works have been commenced, and it is hoped by a vigorous prosecution of the works speedily to render communication between Bristol and South Wales complete.' But it had no powers to do this, its Act mentioning only a steam ferry. The 11½-mile line, broad gauge, was opened 8 September 1863 and worked by the GWR, which joined it at Lawrence Hill. The detached northern section between the pier and a station on the South Wales Railway is often quoted as having been opened on 1 January 1864, but historian John Norris is sceptical. He feels that if trains had not met the ferry from the previous September, the fact would have been on record. He points to a complaint about inadequate motive power in the interim period, which is in the Union company minutes.

All the B & SWU stations were otherwise on Bristol's outskirts: Lawrence Hill, Stapleton Road, Filton, and Pilning—probably on the site of the later Low Level platforms. Ashley Hill was added in 1864.

Ferry sailings were maintained by a contractor using a steamer provided by the railway company, whose trains ran to the end of the piers on both banks. Passengers boarded the ferries from pontoons connected to the piers by stairs and lifts. The pontoons were necessary because of the tremendous tidal range of up to 46ft.

The stone section of New Passage Pier remains—a monument not so much to a forgotten railway, but one that has changed out of all recognition. It began when the GWR absorbed the company in 1868. Doubling took place between Bristol and Narroways Hill Junction, north of Stapleton Road, followed by the rest of the route in 1886, just before the Tunnel opened. It created one of the most interesting stretches of main line in Britain, where levels are split. The separate Up line between Patchway and Pilning was engineered on a much more favourable (and uniform) gradient of 1 in 100, compared with 1 in 68 of the original alignment. A tunnel exactly one mile long, was 243yd shorter than the original, called Patchway Old. The B & SWU route from its divergence with the Tunnel approach to New Passage Pier became redundant when the ferries ceased, but most came back to life fourteen years later when $1\frac{1}{4}$ miles were incorporated into the Pilning–Avonmouth line of 1900 (page 39).

As finally adapted, much of the once humble Union line gained an importance it retains. As D. S. M. Barrie pointed out in the *Railway Magazine* of December 1936:

> Since trains to and from South Wales or Bristol over the South Wales and Bristol Direct line via Badminton also now use large portions of it, this purely local line later assumed an importance far greater than its initial sponsors can have imagined possible.

SEVERN TUNNEL

Curtain-raiser to the Severn Tunnel scheme authorised in 1872 was the South Wales & Great Western Direct Railway, authorised in 1865 from Wootton Bassett to Chepstow—a line of forty-one miles, crossing Oldbury Sands by a bridge or viaduct $2\frac{1}{4}$ miles long and 100ft above high water. After four years the Company was stated to be 'in abeyance,' not having attempted to raise capital.

Inky blackness inside the long tunnel is a poor memorial to a man's vision. Only those who study the history of the Severn Tunnel know that the moving spirit through much personal adversity was Sir Daniel Gooch. Not for him a spectacular memorial like Brunel's Tamar bridge.

A suggestion of Charles Richardson, the GWR resident engineer on the Bristol & South Wales Union, crystalised into an Act of 1872 and work began at Sudbrook on the Monmouthshire shore the following year. Progress was slow, due mainly to the problem of water getting into the workings—not from the Severn, but from fresh water springs which abounded in the area, and from a large natural underground reservoir. In 1879, when construction was well advanced from both banks, water from the reservoir totally flooded the workings and construction came to a halt.

A contract for completion was placed with a London contractor, T. A. Walker (who later built the Manchester Ship Canal). Walker had wide experience of constructing railways under London. He had worked with Sir John Hawkshaw, consulting engineer to the Severn scheme, who was now put in direct charge by Gooch. Richardson was retained as joint engineer, in charge of about 3,600 men.

The works remained flooded until late 1880, but if the GWR directors had any thoughts of abandoning the project in favour of a bridge, their thinking must have been influenced by the Tay Bridge disaster the year before. It occurred only weeks after the opening of the Severn railway bridge, authorised at the same time as the tunnel. But the tunnel was far more ambitious. A bore $4\frac{1}{4}$ miles long was needed to cross the Estuary a little under half that width and to get 30ft beneath the river bed. Stiff gradients, mainly at 1 in 100, were needed. A major pumping operation was—and is—still required to clear twenty million gallons of water a day, and the NUR exempts pumping staff from taking industrial action.

The tunnel scheme involved construction of more than seven miles of line from Pilning Junction to Severn Tunnel Junction, created at a village called Rogiet. On opening of goods 1 September 1886, and to passengers three months later, the tunnel caused the closure of the New Passage ferry. History was repeated when the Aust car ferry was displaced by the Severn road bridge 80 years later.

GREAT WESTERN RAILWAY

In consequence of the destruction of PORTSKEWETT PIER by Fire, arrangements have been made for the following

SERVICE OF TRAINS

BETWEEN

CARDIFF & BRISTOL

Via LYDNEY and SEVERN BRIDGE, to be commenced on WEDNESDAY, MAY 25th.

SPECIAL TIME TABLE OF TRAINS BETWEEN CARDIFF & BRISTOL. (Week Days only.)

UP.		CLASS 1 2 3 A.M.	CLASS 1 2 3 A.M.	CLASS 1 2 3 P.M.	CLASS 1 2 3 P.M.	CLASS 1 2 3 P.M.
New Milford	dep.		6 45	8 40		1 5
Carmarthen Junction	,,		8 30	10 13		2 45
Swansea	,,	7 0	9 45	11 5		3 55
Neath	,,	7 25	10 18	11 35		4 26
Bridgend	,,	8 13	11 10	12 9		5 10
Cardiff	arr.	9 7	12 8	12 39		6 6
				P.M.	P.M.	P.M.
CARDIFF	dep.	9 15	12 50	3 5	6 30
NEWPORT	,,	9 40	1 15	3 30	7 0
PORTSKEWETT	,,	10 3	1 38		7 28
CHEPSTOW	,,	10 13	1 48	3 58	7 38
LYDNEY	,,	10 30	2 5	4 15	8 0
BRISTOL	arr.	11 45	3 10	5 30	9 5

DOWN.			CLASS 1 2 3 A.M.	CLASS 1 2 3 P.M.	CLASS 1 2 3 P.M.	CLASS 1 2 3 P.M.
BRISTOL	-	dep.	8 25	12 45	3 10	7 10
LYDNEY	-	,,	9 30	1 50	4 15	8 20
CHEPSTOW	-	,,	9 47	2 7	4 32	8 37
PORTSKEWETT	-	,,		2 17		
NEWPORT	-	arr.	10 14	2 40	5 0	9 5
CARDIFF	-	,,	10 40	3 5	5 30	9 32
			1 & 2 Exps.		1 & 2 Exps.	1 & 2 Exps.
Cardiff	-	dep.	11 20	3 20	6 7	10 15
			P.M.			
Bridgend	-	,,	12 15	3 53	7 5	10 50
Neath	-	,,	12 59	4 30	7 40	11 27
Swansea	-	arr.	1 30	4 57	8 15	11 55
						A.M.
Carmarthen Junction	-	,,	2 37	5 45	10 0	12 43
New Milford	-	,,	4 12	7 15		2 0

The 10.45 a.m. Train from Portskewett Junction to Cardiff, and the 9.5 a.m. Train from Portskewett to Chepstow, will be discontinued, and the 8.55 a.m. Train from Monmouth to Portskewett will not run beyond Chepstow.

NOTE.—The Fares from CARDIFF and NEWPORT to BRISTOL, and *vice versa*, will be as under :—

First Class. Single - - - - 8s.	First Class, Return - - - - 12s.
Second do. do. - - - - 6s.	Second do. do. - - - - 9s.
Third do. do. - - - - 4s. 6d.	

The Single Fares between Bristol and other Stations will be 3s. First Class. 2s. 6d. Second Class, and 2s. Third Class, *in addition* to the fares hitherto charged via New Passage.

J. GRIERSON,
General Manager.

Paddington, May 24th, 1881.

SOUTH WALES & BRISTOL DIRECT RAILWAY

The Cotswolds were once more scarred, though not unpleasantly, when the GWR embarked on the 33-mile South Wales & Bristol Direct Railway. It was needed and justified for several important reasons and its background is worth examination. It was born amid growing discontent among South Wales industrialists against the GWR's monopoly of the route to London, and they strongly supported a Bill deposited in 1895 for one slightly shorter than that of the GWR, consisting of about 163 miles of new railway costing well over £5 million. It was to run from the Barry Railway at Cardiff, pass north of Newport, cross the Severn at Aust on a spectacular bridge, 3,300yd long, and continue via Thornbury, Malmesbury, Cricklade, Oxford, Chinnor and Bledlow to join the Metropolitan just north of Great Missenden and follow a new route to yet another London terminal. Shortly before the Bill was due in Parliament the promoters reached agreement with the GWR, which went ahead with its Badminton route, one of several short cuts which helped it to get rid of the Great Way Round image.

It reduced Paddington–Newport mileage to 133, which was ten miles less than via Bath, twenty-five miles via Gloucester, and relieved the London–Bristol main line of some congestion over its narrowest and steepest section, that through Bath. Part of this section, though not the worst of it, was used by South Wales trains taking coal to the Royal Navy at Portsmouth and to the Merchant Navy, including the great ocean liners, based at Southampton. It gave Bristol a Paddington route nearly a mile shorter and far better speed-aligned, the 'Badminton route' being laid to nothing steeper than 1 in 300.

Authorised in Victorian days (1896), the Badminton line was not completed until Edwardian times (1903), partly because heavy engineering works including embankments, cuttings, a major tunnel (Sodbury—just over $2\frac{1}{2}$ miles) and several notable viaducts. A marshalling yard was developed at Stoke Gifford, initially ten sidings for 500 wagons, with land available for many more.

The Badminton route was completed in stages: Wootton Bassett to Patchway was opened to goods on 1 January 1903, and Badminton–Patchway five months later. The line opened to passengers on 1 July. The Westerleigh spurs to the Midland's Birmingham line were added on 9 March 1908. Two years later the Badminton route gained another important feeder in the Avonmouth direct line from Stoke Gifford (page 41), and became a junction for a rural branch line when that of the Malmesbury line was switched to Little Somerford in 1933 (page 31). More recently, local passenger services were withdrawn in 1961, although Badminton had a station until 1968. Four years later Bristol Parkway station opened, some months after virtually all the Stoke Gifford Yard had been closed. Subsequently the Badminton line was closed for several months while it was upgraded for High-Speed Trains.

SECONDARY, INDEPENDENT & MINOR BRANCHES

Besides the main lines into South Wales there were two single lines, very much secondary, which fed traffic to and from South Wales while usefully exploiting once-rich ironstone fields. There was the Stratford-upon-Avon & Midland Junction Railway and, further south, a rather more substantially conceived and developed line which was one of the GWR's longest branches.

BANBURY—CHELTENHAM

The 47-mile Cotswold curvacious route was built by three companies over more than three decades, 1855–87. Its promoters saw it more as an important route in a cross-country link between South Wales, the East Midlands and Eastern England, rather than as a purely local line serving well-spaced small towns and villages, often of great beauty and antiquity.

First came the CHIPPING NORTON branch, built as the Oxford Worcester & Wolverhampton responded to pressure from the established and busy market town lying only 4½ easily-conquered miles from its main line. Authority in summer 1854 led to a junction called Chipping Norton being opened just over a year later on 30 August 1855. It became better-known as Kingham in 1909. Although powers were obtained in the name of the OW & W, the Company had separate capital, largely subscribed locally, and effectively its own directors. The company amalgamated with the GWR in 1863.

Meanwhile, the West Midland Railway had played a close part in completing what became the second section of the through route. The BOURTON-ON-THE-WATER Railway. Authorised in 1860, it ran west from Chipping Norton Junction for 6½ miles and opened with only partly-completed stations, 1 March 1862. It was a separate company for many years, although the West Midland/ GWR worked it from opening.

Three-quarters of the route was provided by the BANBURY & CHELTENHAM DIRECT Railway (thirty-three miles), which emerged in 1873 from a number of schemes. The first section, Cheltenham to Bourton-on-the-Water, opened 1 June 1881; the second half was completed six years later [6 April 1887] between Chipping Norton and the Birmingham–Oxford line at King's Sutton (also spelt without apostrophe), 3½ miles south of Banbury and only a mile north of what became Aynho Junction.

Traffic was enhanced after the Midland & South Western Junction Railway was completed from Cirencester to the Direct line at Andoversford in 1891. The six miles west to Cheltenham (Lansdown Junction) were doubled in 1900 and the through route was improved by the Hatherley Curve (west–south) at Cheltenham and the Kingham Avoiding Line, both opened 8 January 1906. The Avoiding Line improved operating on the OW & W and the Direct Line by avoiding the need for through trains to reverse.

The importance of the Direct line was that it ran through the middle of the main Oxfordshire ironstone field, developed in the middle of the last century as supplies because exhausted in other parts of Britain. A string of mines close to the line were connected by standard or narrow gauge systems. The ore went to South Wales, Staffordshire, and North Wales, the Brymbo Steel Company having a siding at Hook Norton. Mining declined after World War II, either because reserves were exhausted or because it became too costly to transport the ironstone to steelworks in several areas where ore was close to hand.

In the event, passenger services went first, beginning in 1939 with the most famous, the Ports to Ports Express or, as it was noted in GWR public timetables 'Through Train Cardiff to Newcastle-on-Tyne. Luncheon Car Train.' It was introduced after completion of the Kingham Direct Line and Hatherley South curve, Cheltenham South & Leckhampton being its only regular booked stop between

Banbury and Gloucester. Cheltenham South was rebuilt so that the Spa would have a station befitting a titled train.

The route also carried a Cheltenham–Paddington through coach service, slipped at Kingham for many years to provide the fastest service between Spa and Metropolis. First local trains to be withdrawn were those between King's Sutton and Chipping Norton, in 1951. The Kingham–Cheltenham (St. James') passenger service was withdrawn in October 1961 and as M & SWJ passenger services had been withdrawn the previous year, Cheltenham (Lansdown Junction)–Bourton-on-the-Water was closed completely. Kingham–Chipping Norton passenger services were withdrawn in December 1962 and with them went the Chipping Norton–Hook Norton freight section. Hook Norton–Adderbury went eleven months later, the 1½-mile eastern branch stub between Adderbury and King's Sutton surviving until 1969 because of limited ironstone traffic. The two freight stubs from Kingham–to Bourton and Chipping Norton had been closed in September 1964.

THE STRATFORD-UPON-AVON & MIDLAND JUNCTION RAILWAY

Born of two separate schemes to get Northamptonshire iron-ore to South Wales, the Stratford-upon-Avon & Midland Junction had a career of mixed fortunes, with two world wars providing the greatest traffic stimulations.

Even its creation was protracted. The main stem of thirty-three miles was provided by the East & West Junction Railway of 1864, authorised to the GWR Stratford branch at Stratford from a junction near Towcester on the Northampton & Banbury Junction.

Bedevilled by cash shortage, the first section of the SMJ was not ready until 1 June 1871, six miles from Fenny Compton (siding connection with the Birmingham & Oxford) west to Kineton. The contractor was Thomas Russell Crampton, better known for his locomotives. The rest of the line opened two years later, and after originally using the GWR station at Stratford it opened its own in 1876. It was renamed Stratford (Old Town) as late as 1952.

The SMJ 'route' was extended by another small and independent company, the Evesham, Redditch & Stratford-upon-Avon Junction Railway, authorised in 1873, and opened 2 June 1879. The eight miles from Broom, Birmingham & Gloucester Loop, to Stratford filled a blank on the railway map originally considered for conquering by the London Worcester & South Wales Railway in 1865, which was also to have served Stratford.

Stratford's importance as a railway centre increased minimally when the SMJ was created in 1908 by the amalgamation of the two companies, together with the Stratford-upon-Avon Towcester & Midland Junction Railway. The offices of the new company were in the modest station buildings at Stratford. They have been long demolished, but the low brick viaduct across the Avon close-by survived.

The amalgamation was a financial salvage following lack of interest in purchasing the route shown by the Midland, LNWR, GWR and even the ever-ambitious Great Central. Yet all exploited the SMJ for their own ends, although the GWR had limited access, confined to spurs and SMJ stations at Stratford and Fenny Compton.

For its part, the Midland ran for thirty years (1883–1913) almost daily goods trains between Bristol and St. Pancras (Somers Town) via Broom and Olney. In later years, banana specials used the route.

The LNWR got to Stratford with Shakespeare specials over the SMJ, while the GCR used the line to develop fierce competition with the GWR, and a pawn in a vain attempt to reach Birmingham (Vol 7).

After Grouping, the LMS tried to exploit the Shakespeare image and attract lucrative tourist traffic by opening the Welcombe Hotel at Stratford in 1931. The following year guests arrived via the station goods yard—aboard a road-rail bus, which changed wheels there during a much-publicised but unsuccessful and short-lived experiment.

More life-giving to the line was the opening of a south curve at Broom in 1942 as the line was developed to keep heavy wartime freight clear of the West Midlands.

Afterwards, economies were not long in coming. Stratford–Broom was closed to passengers by the LMS in 1947 and BR completed the passenger economies by withdrawing the Stratford–Blisworth service in 1952. The Mayor gave a civic send-off to the last train to Blisworth, riding on the footplate to Ettington, where that station had an average of one passenger a *month*.

But the line was far from freight-extinct and it had one further improvement in 1960 when it was chosen as a route for South Wales ironstone from Banbury. The Fenny Compton connection was reversed to allow through running via the Birmingham & Oxford and a new spur was put in at Stratford for trains to reach Honeybourne. The section to Broom was closed completely.

Despite the changes, the traffic dwindled and the line was closed officially in 1965, apart from one length. East of Fenny Compton, the SMJ history falls into Volume 9, but the boundary is convenient (for once!) for it marks the start of the surviving stretch, which runs to an Army ammunition depot opened in 1940 at Burton Dassett. It was transferred to the Ministry of Defence in 1971.

Burton Dassett was once the junction for the EDGE HILL LIGHT RAILWAY, one of the shortest-lived standard gauge railways on record. It was built under a Light Railway Order of 1919 to tap an isolated ironstone crop on top of the Burton Hills, very much with the backing of the SMJ.

It ran just over two miles to the foot of the hills and climbed them on a long rope-worked incline. It closed in 1925 after only five years, but its remains were in situ for years. An RCTS member who visited the route on 12 August 1938 afterwards wrote in *The Railway Observer* 'The state of affairs at the Edge Hill end of the line is amazing, as everything is still practically in the same state as when work was suspended nearly 18 years ago.'

Much of the trackbed disappeared when the Kineton ammunition depot was developed from 1940, although in 1942 two former Terrier tank engines of the London Brighton & South Coast were still lying derelict—just where they had run out of steam nearly two decades earlier.

For years Kemble's railway importance rested partly on branches to the busy market towns of CIRENCESTER and TETBURY which served a wide area of the Cotswolds. Cirencester became a branch terminal in 1845 when the C & GWU was extended to Cheltenham, and the 4¼ miles from Kemble were down-graded. The branch achieved national prominence when steam was replaced in February 1959 by lightweight four-wheel diesel railbuses on this and the Tetbury branch. They attracted extra passengers but not enough, and they were withdrawn after five years. The Tetbury branch then closed completely; Cirencester freight lingered for 18 months. Cirencester Town's historic building is incorporated into a bus station, the local council having been refused permission to demolish the listed structure.

The Tetbury branch, three miles longer than its neighbour, had a much shorter life, not opening until 2 December 1889, having been created in the wake of a number of unsuccessful schemes. They included one for another GWR route to South Wales, which was to be created by extending the Fairford branch to Cirencester, doubling the Kemble branch and using the Cheltenham line to Stonehouse, which was to be the starting point for a new line.

CHAPTER VI

The West Midland Railway

The West Midland Railway was a company that cast its net far and wide for just three years at a time when a number of companies, big and small, were either authorised or being built. The Oxford Worcester & Wolverhampton changed its name to the West Midland Railway on 1 July 1860, and absorbed the unconnected Newport Abergavenny & Hereford and Worcester & Hereford companies. The WMR was leased to the GWR on 1 July 1861 and amalgamated with it from 1 August 1863. During its existence, the West Midland had agreements either to lease or work six branches, not opened when the company was formed. They were the Witney and Bourton-on-the-Water, the Severn Valley, and Stourbridge Railway (Vol 7) and two lines that stemmed from the Severn Valley–the Tenbury & Bewdley and Much Wenlock & Severn Junction.

The 1860 total of seventy-seven miles of leased or worked lines grew when the WMR took-over working the Leominster & Kington and Coleford Monmouth Usk & Pontypool. The Tenbury & Bewdley did not open in its lifetime, but by 1863 the WMR also owned about 200 miles of railway.

OXFORD WORCESTER & WOLVERHAMPTON

Despite the change that came over railways in the following century its spirit never quite disappeared, for BR service timetables continued to sub-title the OW & W as the 'West Midland Section.'

Leaving Oxford for Worcester on a sunny day, the Inter-City

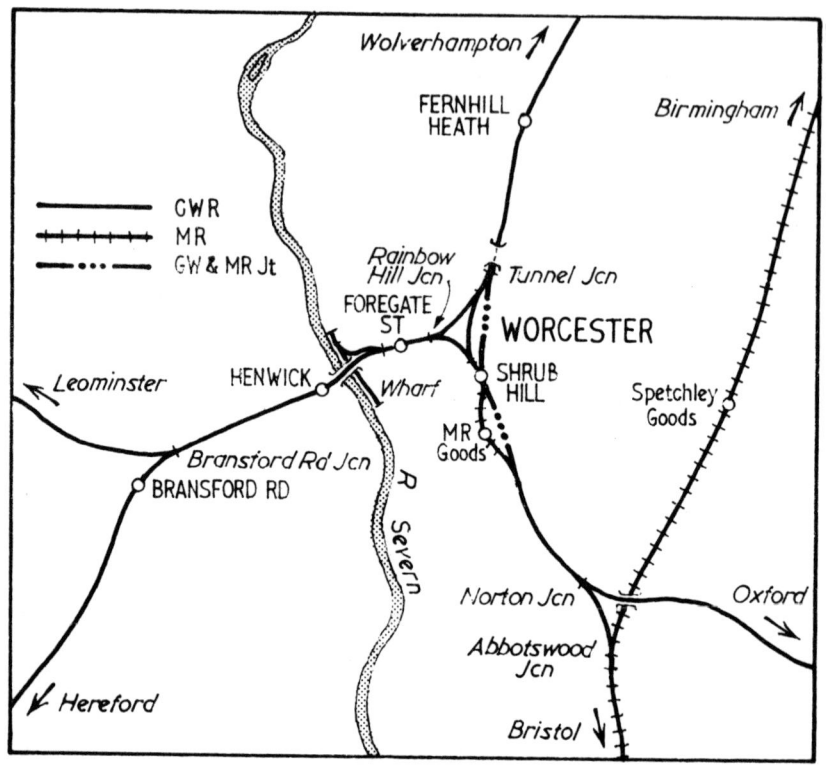

traveller faces the prospect of a run through some of the quietest and loveliest countryside of England. The rural character of the Cotswold Line, as it is now known, gives no hint of its stormy birth, mainly to try to satisfy the demands of industrialists when the Industrial Revolution was raging through the West Midlands nearly a century and a half ago.

The Oxford Worcester & Wolverhampton Railway was conceived to smash the London & Birmingham's monopoly of a route between the Black Country, London, and other southern ports through which Midland manufacturers dispatched their goods all over the world. Promoted by Worcestershire men, the OW & W Act of 4 August 1845 (amid the Railway Mania) was for an 89-mile route from the broad gauge Oxford branch of the GWR (with which its gauge was to be held in keeping), via Worcester, Droitwich, Kidderminster and Dudley to the Grand Junction at Wolverhampton (Vol 7).

The Act followed a successful battle against the London &
Birmingham, which proposed a line from Tring to Wolverhampton
through the Vale of Evesham. There were suspicions that the
London & Birmingham would not build a route to duplicate its own,
while Parliament preferred the Worcester scheme because of the
company being independent and its plans to open up a much bigger
area of countryside.

The OW & W was supported by the GWR, and with Brunel as
engineer Paddington saw itself poised to extend the broad gauge
deep into the West Midlands—a long stride towards its dream of
broad gauge to the Mersey. But there were troubles from the start
and they led to the OW & W having one of the stormiest histories
of any English company. The troubles began after a wild estimate
of cost by Brunel. He had to increase his original figure of £1,000,000
by 2½ times, and that led to shareholders arguing with the GWR on
how much return they were to be guaranteed. Money soon ran out
and there was no income because the whole route had been started
at once and not even a short section was nearing completion. A
further blow which angered shareholders was the GWR's purchase
of the Birmingham & Oxford and Birmingham Wolverhampton
& Dudley in 1848, which provided a route to the south shorter and
more direct than their own via Worcester was to be.

The first two short sections of the OW & W opened 5 October
1850 were narrow gauge: they made up the four miles between
Worcester and the Midland's Birmingham & Gloucester route at
Abbot's Wood Junction. They were followed on 18 February 1852
by completion of the Worcester Loop: Worcester–Droitwich–Stoke
Works Junction (9¼ miles). This was an important stretch, for
Droitwich was growing as a spa, much favoured by the Victorians.
It was to lead the GWR to erect 'an excellent Railway Station in
1899 with a special view to the comfort of invalids.' The quote is
from Darlington's Handbook *The Severn Valley*, which also stated
that the great rock-salt beds at Stoke Prior were discovered in 1828.

By 1852, construction of the OW & W was well under way and
1 May saw the opening of two more sections—Droitwich–Stourbridge (16½ miles), and ten miles between Worcester (Norton) and
Evesham—at a time when a lease to the LNWR was being considered, together with an extension from Wolvercot, near Oxford,
to the Hounslow branch of the London & South Western via Thame,

Wycombe and Uxbridge. Sir Morton Peto, MP, referred to it at the Evesham celebrations, stating that the London extension 'should and must be made.' Fellow MPs thought otherwise, and the LNWR lease was also rejected.

Evesham remained a terminus only a short time, for the 40½ miles through to Wolvercot were completed single line 4 June 1853, doubling following over the next five years. The stretch included Mickleton (later Campden) near the line's summit above Honeybourne, where Brunel massed 2,000 navvies when a sacked contractor refused to leave the site. Troops were called from Coventry, but the contractor retreated after a minor skirmish and they were not needed. The line was completed in the Black Country, advancing from Stourbridge Junction to Dudley and Bilston and reaching Wolverhampton in April 1854, an extension to the LNWR at Bushbury Junction following in July (Vol. 7).

Operationally, the infant Worcester route got a terrible reputation, especially for passenger services. Mainly because of unreliable locomotives, it took some time to shake off its nickname, endowed locally, of the Old Worse and Worse.

YARNTON LOOP

The Worcester route was joined to the LNWR at its southern end, following the opening by Euston on 1 April 1854 of a 1½-mile spur from the Buckinghamshire Railway north of Oxford, to Yarnton. It was a Parliamentary sop to Euston when its ambitions for a Tring–Oxford branch were thwarted. The GWR was angered when through Euston–Worcester trains were introduced via Bletchley and Yarnton, but relations later improved and excursions began between the Worcester route and the South-East via Reading, where trains used a new spur to the GWR standard gauge metals.

Euston placed high value on the Yarnton Loop as forging a through route between South Wales, the West Midlands and Eastern England, while regarding the capital cost of £36,000 as high. The chairman, Lord Anson, explained to shareholders: 'It is only two miles in length, but being in the immediately vicinity of Oxford, where land is of higher value than the agricultural districts, the expense is somewhat enhanced.'

Plate 17
Above: Few photographers could resist climbing the hill overlooking Worcester shed. The 1904 scene shows the yard well filled with shunting tanks and 0-6-0 goods engines, and wagons of locomotive coal, including several of the Midland Railway. (*L&GRP, courtesy David & Charles*)

Plate 18
Below: Monmouth (Troy), with an auto train standing just outside the station on the line to Lydbrook Junction, while the pannier tank with open wagons loaded with steel bar turns off to the right towards Lydney. (*HMRS collection*)

Plate 19
Above: Epitome of GWR glory. Kidderminster in the 1880s. Outside frame 2-4-0 with highly polished dome and safety valve cover on an up train. (*Wyre Forest District Council Museum & Art Gallery Service*)

Plate 20
Below: Bridgnorth before preservation! The journey between Shrewsbury and Kidderminster was sufficiently lengthy to warrant corridor coaches. (*W. A. Camwell*)

THE COTSWOLD LINE TODAY

The OW & W is a very different railway from what it was only some twenty years ago, being virtually fragmented into three passenger service sections. The major loss of freight traffic occurred when BR abandoned plans on which it had spent heavily (including constructing the Bletchley flyover) to develop an East Anglia–South Wales route via Honeybourne and Cheltenham. It was to keep freight clear of the then congested London lines. The Yarnton Loop, an essential part of the line, closed in 1965 and the next year began with the closure of a number of wayside stations as a compromise to BR's own proposals for the withdrawal of Oxford–Worcester local services. Among the stations to go was Adlestrop, but as *The Observer* pointed out when the plan was announced by the Minister of Transport, Tom Fraser, that Edward Thomas wrote its epitaph fifty years before Beeching. With Thomas' poem, Adlestrop lives on.

The Observer described the Worcester–Oxford line has having stations with evocative names and infrequent times and also cited Littleton & Badsey. Not placed in that category was Kingham, which had ceased to be a junction for Cheltenham in October 1962 and for Chipping Norton only two months later. It survived on Government insistence and in 1975 its imposing buildings were replaced by a modified Southern Region design, 'sympathetic to the mellow Cotswold Countryside,' stated *Rail News*.

Charlbury retained its waiting room fire after commuters petitioned against its replacement by an electric one. BR chairman Sir Peter Parker was among protesters.

In the early 1970s, forty of the fifty-seven miles between Oxford and Worcester were singled. In 1978, Sir Peter welcomed the formation of the Cotswold Line Promotion Group. It got Honeybourne station (closed 1969) re-opened in 1981, and an improved service provided for Pershore. The plan may be scrapped if the economic situation improves, but as a curtain raiser some Paddington–Worcester services were diverted via Gloucester, while the Swindon–Gloucester–Cheltenham diesel multiple-unit service was extended to Worcester. These trains give HST connections at Swindon. Worcester NUR members described the Inter-City switch as a 'bombshell', fearing the economy could lead to closure of the Cotswold Line. BR has announced that Oxford–Worcester will lose its Inter-City status from May 1982, when passengers will have to change at Oxford. The aim is to replace locomotive hauled expresses with light-weight units to avoid spending £1,500,000 on track renewal.

Worcester is not a major generator of commuter traffic, for its population has grown only slowly. It has yet to double the 40,000 total of almost a century ago. But it is the hub of a large and prosperous area, and a radical change in its railway pattern is possible under plans still being discussed after several years for a park-and-ride station on the outskirts, serving at different levels the Birmingham and Gloucester and Oxford lines. It would help to keep traffic away from the city centre, allowing Shrub Hill station to close and Foregate Street, more centrally situated, to be developed.

Local NUR members were immediately critical because of the proposed loss of Shrub Hill. They work in a city where economies have continued since Nationalisation. Among the earliest was downgrading the former GWR divisional locomotive headquarters at the end of 1951, when the casual repair shop became a locomotive concentration depot. Two branches have gone: the short 'Butts' branch to the riverside racecourse, authorised in 1859 and believed to have opened in 1862, which ran from Foregate Street to a station tucked at low level beside the W & H viaduct across the Severn. The line closed in February 1957. Seven years later came closure of the half-mile 'Vinegar branch', so called because it served the vinegar and cider works of Hill Evans & Company from summer 1872. It attracted enthusiast interest because motorists had to obey GWR standard lower-quadrant signals which guarded two crossings over busy roads. The signals were worked from ground frames. The branch also crossed a factory approach road on a lifting bascule bridge.

More recently, full wagon-load traffic handled by Worcester NCL depot was transferred to Birmingham (Lawley Street) in 1974, and Shrub Hill station lost its parcel post traffic from May 1976. Amid NUR protest, all Worcester's parcel post (with the exception of Moreton-in-Marsh) was switched to road between Birmingham and local post offices.

For years Kidderminster did much to project rail facilities as one of its assets, offering manufacturers 'speedy and efficient transport by the GW and LMS Railways at moderate freight charges.' While world-famous for its carpets, it has continually sought to broaden its industrial base, and the town of today has some 50,000 inhabitants—a population growth of around 20,000 in forty years. Until 1974 it had a black-and-white country house style station (reputedly intended originally for Stratford-upon-Avon). Its replacement was a single-storey utilitarian building. It lies on the fringe of the town centre, a fact that the Western Region recorded in timetables, stating that there was a bus service between the station and town centre.

WORCESTER AND HEREFORD RAILWAY

The ridge of the Malvern Hills was once an unforgettable place from which to watch a steam train's progress through a chequerwork of green fields.

A wispy trail of white steam marked its approach from Worcester to Malvern Link station, Great Malvern (a masterpiece of Victorian elegance remembered today in such features as the ornate cast-iron capitals), and Malvern Wells station, just before disappearing below one's feet into one of the deepest tunnels in Britain. Out again, the trail resumed west towards Ledbury and Hereford.

Except for the type of train, the line has changed little since it linked the Cathedral Cities—a fact acknowledged by The Cathedrals Express. Elgar knew this line well when he lived at Malvern and Hereford. The Worcester & Hereford, authorised in 1853, four years ahead of his birth, came in the wake of several schemes aimed at creating a through route between the Midlands and South Wales by removing the $29\frac{3}{4}$-mile gap between the OW & W at Worcester and the Newport Abergavenny & Hereford. While it was under construction, the Midland and LNWR projected a line in 1851 from the OW & W just south of Shrub Hill station, Worcester, crossing the Severn near Diglis and passing north of the Malvern Hills. The Bill failed, partly because of vocal opposition from Great Malvern and Ledbury, angry that the railway was to avoid them. The W & H of 1853 was a purely local company and its route came to grips with obstacles the earlier scheme avoided, squeezing through a built-up area of Worcester and piercing the Malvern Hills and a ridge near Ledbury with long tunnels.

These engineering works proved to be beyond the financial resources of the small company, which found itself unable to adopt the usual course of getting help from its wealthier neighbours. It was not until 1858 that Parliament gave subscription powers to the OW & W, NA & H and Midland. In *Bradshaw's Manual* of 1859 (which I bought while attending the Three Choirs Festival in Worcester 110 years later) the company reported that 'The works, under contract with Mr Brassey, are in active progress. The portion between Worcester and Malvern may be opened in the summer of this year.' The statement was optimistic, for while the six miles between Malvern Link and Henwick opened that July, completing the Severn crossing by bridge and viaduct over low-lying land took until 17 May 1860, followed eight days later by the two miles between Malvern Link and Malvern Wells. Malvern trains from Shrub Hill had to reverse at Tunnel Junction until the quarter-mile spur to Rainbow Hill Junction opened 25 July.

Malvern Tunnel caused further delay and it was not until 13 September 1861 that the last eighteen miles were completed to the Shrewsbury & Hereford at Shelwick Junction, north of Hereford. The stretch included Ledbury Tunnel of 1,323 yd. Malvern Tunnel proved so troublesome that the 1,567 yd single bore was replaced in 1926 by one 22 yd longer, but on a slightly easier gradient. It still presents operating problems: in 1975 NUR guards protested about new instructions for working diesel multiple-units through the tunnel.

For years, Worcester–Hereford remained Inter-City more in name than anything else, since all trains stopped at the four intermediate stations—Malvern Link, Great Malvern, Colwall and Ledbury—and BR timetables showed the fastest Hereford–Paddington route was via Newport. Much of the Worcester line was singled after BR selected the Birmingham–Bristol main line in its 1965 report on *The Development of the major Railway Trunk Routes*. That was then carrying 90,000 tons of freight a week, 20,000 tons more than the W & H, a route which offered no mileage advantage between the West Midlands and South Wales since they are roughly the same.

At Worcester, Rainbow Hill Junction was abolished in the 1970s leaving the twin tracks through Foregate Street station to Henwick as separated single lines. Casual passengers using the station can get confused by finding Birmingham and Hereford trains in both directions using the same platform. In summer 1981, fast Birmingham–Worcester (Foregate Street)–Malvern (hourly) and Hereford (two-hourly) diesel multiple unit services began with 'substantial help and encouragement from Hereford & Worcester County Council,' to quote handbills. They run non-stop to New Street from Bromsgrove (where they sometimes take a banker). Malvern is under an hour's rail journey from Birmingham; Worcester, less than 40 minutes—timings not possible via the Kidderminster route, seven miles longer.

The Worcester & Hereford of today is very much a withered trunk which has lost its branches, for those to Bromyard and Leominster, Ashchurch and Gloucester have long closed. Their passenger services were well advertised in *The Malvern Railway Guide*, whose orange-covered edition for May 1940 (while the Germans were over-running France) warned 'fares liable to 10% increase.'

NEWPORT ABERGAVENNY & HEREFORD

Hereford was one of the last towns of comparable size to get railways. Yet between 1852–64 it got four; the first two which formed a major trunk route were by far the most important. The Newport Abergavenny & Hereford was incorporated on 3 August 1846—the same day as the Shrewsbury & Hereford, which opened ahead of it—to goods 30 July 1852 and to passengers 6 December 1853. The Newport line was not completed until 16 January 1854.

The impact on Hereford was immediate. Murray's *Hand Book for Travellers in South Wales and its Borders* noted in 1860 (five years after the Hereford Ross & Gloucester had opened, but four years ahead of the Brecon line): 'Hereford has always maintained a staid and quiet dignity which contrasts pleasantly with the bustle and restlessness of a manufacturing town; although within the last few years the convergence of 3 or 4 lines of railway has imparted to its streets a degree of animation to which it was long a stranger.'

While Hereford may have had a useful web of lines, stations were badly sited on opposite sides of the city centre: Barr's Court to the east, Barton and Moorfields to the west.

When promoters first thought about lines along the Border, Hereford's population was small. It did not reach 20,000 until the early 1880s. Since then it has more than doubled: the 1971 figure was 47,000.

The NA & H grew out of a rejected grand scheme of the Railway Mania for the Welsh Midland Railway, planned from the Birmingham & Gloucester near Worcester via Hereford and Brecon to Merthyr, where the Taff Vale was to provide access to Cardiff, Swansea and other places. Because of the monetary crisis after the Mania, construction of the NA & H was suspended from 1847–1851.

In 1851 the LNWR began a drive into South Wales and although Parliament refused sale of the NA & H to Euston, the LNWR worked the line from its opening. The Border route had been forged together by a mile of single line north from the NA & H headquarters at Barton Station, Hereford, to Barton (later Barr's Court) Junction, built by the NA & H using powers of the Worcester & Hereford, then far from complete.

The NA & H ran south thirty-three miles to Coedygric Junction, about a mile south of Pontypool Road, of the Monmouthshire Railway. It was another of Euston's allies and the LNWR worked trains to the MR terminus, a small station called Newport Mill Street, close to the docks and at a lower level than the South Wales main line. The arrangement lasted until the opening of the Pontypool Caerleon & Newport Railway (Vol 12) in 1874. Promoted by the GWR it ran nine miles from Pontypool South Junction on the opposite (east) bank of the tidal Usk to join the South Wales main line at the sharp triangular Maindee Junction, within sight of Newport (High Street) Station.

Euston's presence on the Welsh border led to many often bitter political clashes, one of the early ones being with the NA & H when Captain Mark Huish, Euston's ruthless visionary, attempted to gain complete control of the Border lines. In this instance, he refused acceptance of Merseyside (Birkenhead) and Midlands (Wolverhampton) traffic not routed via Stafford, a ridiculously roundabout route.

The NA & H switched working from Euston to Thomas Brassey, who as contractor was then working the S & H, and a second Parliamentary application to sell the Abergavenny company to Euston was abandoned. Lean days with many operating problems followed before the West Midland Railway was formed and the company faced a more settled future. By then, the NA & H had twenty-five locomotives based on Barton, which subsequently became a GWR depot, the offices being the original headquarters of the NA & H.

Realising that it would not achieve full potential until the Worcester & Hereford opened, the Abergavenny took powers to subscribe up to £37,500, while looking to the Midland Railway to take the main development initiative.

Despite setbacks in relations with the NA & H, Euston consolidated its position in South Wales as a result of the outcome of two proposals to take railways across the mountains from the Newport route, put forward in 1858.

The NA & H supported the Breconshire Railway & Canal Company which aspired among other things to convert its canal from Abergavenny to Brecon into a railway and adapt part of the Llanvihangel Railway, incorporated in 1811. This was one of three tramroads taken-over by the NA & H. The others were the Grosmont Railway of 1812 and the Hereford Railway of 1826. Together they formed a tramroad between Abergavenny and Hereford. It ran more-or-less beside the main road, while the NA & H cut a new route.

A rival scheme to the Breconshire showed more promise by following a better and more direct route into the Valleys. The Merthyr Tredegar & Abergavenny was supported by the South Wales industrialist and railway pioneer Crawshay Bailey.

The Merthyr Company was authorised in 1859 once the NA & H withdrew support for the Breconshire scheme and the Act allowed transfer from the NA & H of part of the Llanvihangel Railway. Euston secured control of the Merthyr route after the West Midland Railway (as successor to the NA & H) was refused leasing sanction by Parliament. Completed between 1862–73 the Heads of the Valleys line, as it was also known, made the Hereford line a busy route and Abergavenny became a bustling little railway centre with the distinctive flavours of the Great Western and LNWR, always full of interest.

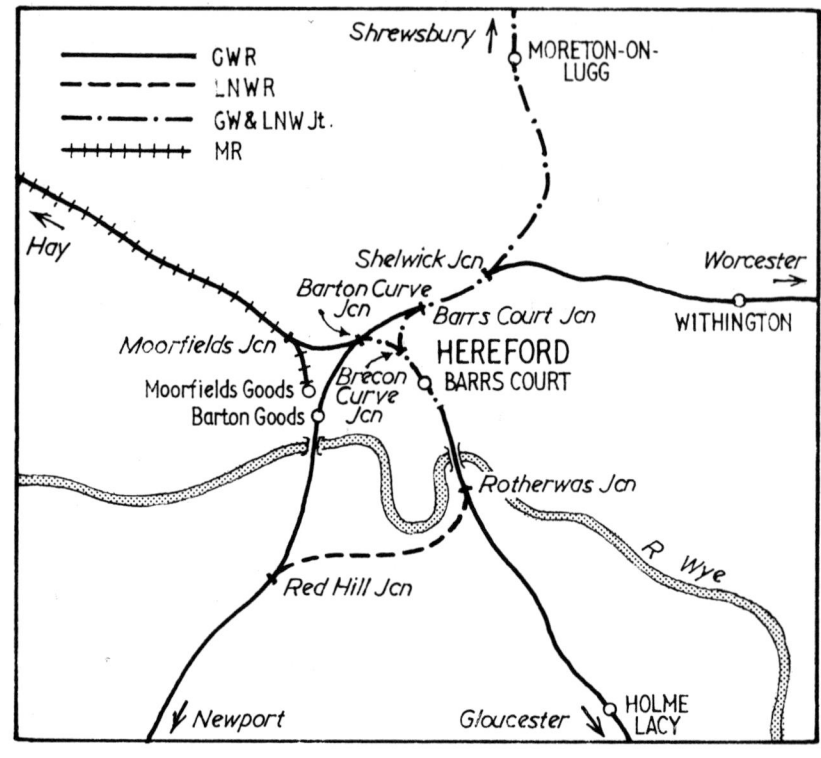

HEREFORD: LNWR LOOP AND OTHER DEVELOPMENTS

Major improvements were made at Hereford in 1866 when the
LNWR built a two-mile link to eliminate reversal of north-south
through trains and standard gauge GWR services at Barr's Court
station. The link, from the GWR at Rotherwas, $1\frac{1}{4}$ miles south of
Barr's Court on Hereford Ross & Gloucester line to the NA & H
at Red Hill, was opened 16 July. The GWR assisted by mixing and
doubling the HR & G section and the conversion (one of the earliest
it carried out) ended the broad gauge intrusion locally. Greater
benefit was to be enjoyed when North-to-West through services were
introduced when the Severn Tunnel opened twenty years later.

Hereford's passenger services gained further coherence from
2 January 1893 through the switching of the Midland's Swansea
and Brecon stopping trains (there was never an express service)
from Barton station to Barr's Court. It was made possible by the

S & H opening a quarter-mile Brecon curve and a connection to the HH & B near Moorfields. Barton, which then closed to passengers, had the distinction like Stoke Works passenger station, of being owned by one company (GWR) while used only by trains of another. Barton survived for goods until 1979.

NA & H: PRESENT DAY

Few main lines have lost so much through traffic and so many feeder branches and yet survive as healthily as Shrewsbury–Newport. Despite switching North-to-West expresses via Gloucester in 1970, it is one of BR's most important secondary routes, not least because it is the only surviving rail link between North and South Wales. Politicians and local authorities constantly press for improved and faster services to those of 1970s which for passengers between North Wales and Cardiff involve changes at either Chester, Crewe or Shrewsbury to the Crewe–Cardiff through trains which make a number of closely-timed long-distance connections at Shrewsbury, Hereford and Newport (Paddington HST).

So far as Hereford–Newport is concerned, the route is now virtually a through rather than local one. Only Abergavenny and Pontypool (a once impressive junction demoted to a bus stop) remain. Cwmbran New Town, despite a population of over 50,000 is a place through which the railway runs rather than serves.

At Hereford, a dmu servicing area has been established at Barr's Court station, now simply 'Hereford'. Signalling and track layout improvements of the early 1970s included an extra southbound freight line through the station. The Red Hill–Barton section of the 'avoiding line' closed in 1967, but Brecon Curve junction survives to provide access to a power station and several firms including H.P. Bulmer, whose railway centre was built after the company obtained authority in 1968 to restore and run No 6000 *King George V*, six years after it was withdrawn for preservation.

OW & W: TRAMWAYS AND BRANCHES

Stratford-upon-Avon is not especially noted for its railway attractions, but it is pleasant to stroll across the red-brick pedestrian bridge close to the Clopton Road bridge. This once carried the STRAT-FORD & MORETON RAILWAY to the canal basin, where goods to and from the Midlands were transhipped. The horse-drawn tramway—for the railway was nothing more—was a remnant of an 1819–20 scheme by the pioneer railway surveyor William James, who was closely associated with the Stratford Canal. He surveyed a London–Stratford railway, (planning Stratford as a canal tranship for the Midlands), via Moreton-in-Marsh, Oxford, Thame and Uxbridge. Collieries near Coventry owned by James were to be linked by a branch from Shipston-on-Stour.

James lacked money, but his initiative led Lord Redesdale, a landowner, to help promote a 16-mile tramway between Stratford and Moreton. Authorised in May 1821 (only six weeks after the Stockton & Darlington Railway), the 4ft 0in gauge tramway opened 5 September 1826 when a new market was completed at Moreton, centre of a large farming area. A branch to Shipston authorised in 1833, opened 11 February 1836, and the OW & W's Incorporation Act of 1845 included a clause for permanent lease. Later Brunel reported that major engineering works would be needed if locomotives were to be used. Much of the tramway became redundant when the OW & W Honeybourne–Stratford branch of 1859 opened. North of Longdon Road, the nearest point to Shipston, it was little used; probably never after 1904. The section was lifted during World War I and legally abandoned in 1928.

But Shipston managed to retain its tramway, and local opposition forced the West Midland Railway to delete closure powers from an 1862 Bill. Horse-drawn still, the tramway survived long enough for the GWR to take an interest in 1882 when it got authority to up-grade the branch from Moreton, using as much of the tramway as possible. A further Act was obtained in 1884 to allow steam opera-tion, prohibited in the original Act.

The SHIPSTON-ON-STOUR branch, of almost nine miles, was opened 1 July 1889 with mixed trains and known officially as a Locomotive Tramway. Slow speeds made it vulnerable to bus competition and passenger trains were withdrawn in 1929, the GWR experimentally substituting buses for several months. They had little more success in attracting traffic because local people travelled to Stratford rather than Moreton.

One engine in steam freight operation continued through another three decades until official closure on 2 May 1960. A section of *Waterways to Stratford* by Charles Hadfield and John Norris is devoted to the tramways' history. As John Norris noted: 'The cuttings and embankments remain as a memorial to those who embarked on a project which might have become one of the trunk lines of communication in the country.'

The Yarnton–FAIRFORD branch was promoted in the wake of failed grandiose schemes, including the $37\frac{1}{2}$-mile Cheltenham & Oxford Railway, authorised 1847 with a short branch to the busy country town of Witney. That was a year after the OW & W had been authorised to build an 8-mile Yarnton–Witney branch. That, too, came to nought and it was not until the Witney Railway, incorporated in 1859, opened from Yarnton on 14 November 1861 to passengers and coal traffic (and fully to goods the following March) that railways served this sparsely-populated corner of Cotswold. Soon there was clamour for an extension but the East Gloucester Railway of 1864 was over-ambitious: a Cheltenham–Fairford–Witney and Lechlade branch. Of those authorised fifty miles, only Witney–Fairford (fourteen miles) were built. Completed 15 January 1873, the EGR and Witney Railway were GWR-absorbed in 1890. The branch closed throughout to passengers and completely from Witney to Fairford on 16 June 1962, the remainder following 2 November 1970. For years there was talk by the GWR and the Midland & South Western Junction of bridging the nine miles between Fairford and Cirencester, especially during the high unemployment years between the World Wars, but the GWR was never enthusiastic, considering such an extension would merely be a case of competing with its existing lines. Paul Jennings remembered the branch affectionately in *Just a Few Lines*. Rightly so.

Of rather more substance and length was the $39\frac{1}{2}$-mile SEVERN VALLEY RAILWAY, authorised between Shrewsbury and the OW & W at Hartlebury in 1853, but not opened until 1 February 1862, the Oxford company having leased the line from the original private one in 1857. It was a lovely line, but often inconvenient as I realised on being posted to RAF Bridgnorth in 1948. Only once did I use the line, to arrive from Shrewsbury by streamlined diesel railcar, ex-GWR by just a few months. After that, I always caught a bus to and from Wolverhampton and used inter-city (although it was not known by that title then). The Severn Valley was well endowed with railways considering its sparse population, and on 1 June 1878 the GWR further grafted the small, pretty town of Bewdley onto its map with a three-mile branch from Kidderminster, strengthening ties between the Black Country and its rural surrounds. The short stretch remains important, as the life-link between BR and the preservationists of the 12½ miles north to Bridgnorth who have never looked back since a rescue operation amid much hustle (and BR co-operation) after Shrewsbury–Bewdley passenger services were withdrawn in September 1963. Those from Bewdley to Kidderminster and Hartlebury (second class only, Worcester connections) ran until January 1970.

Bewdley–Stourport closed completely, but some of the line was bought by the SVR in 1979. Stourport–Hartlebury followed in 1980. Such is the success of the preserved line that BR began linking Kidderminster and Bewdley with limited connecting services in 1979. The preserved line helps to keep Bridgnorth's economy buoyant, and provides extra passengers for the CASTLE HILL FUNICULAR—a local delight since 7 July 1892, and since closure of the Clifton Rocks Railway at Bristol in 1934, Britain's only inland cliff railway. Bridgnorth would have had another Paddington link if the 1905 Wombourn–Bridgnorth line (Vol 7) had been built. As it was, the GWR branch between Shrewsbury and Worcester was busy for many years. Buildwas, with its split-level station for Wellington–Craven Arms trains, had a staff of ten at Grouping—twice the number of railwaymen at Berrington and Cressage, intermediate stations nearer Shrewsbury. Yet Buildwas passenger, parcel and goods receipts were less than half those of its neighbours.

CHAPTER VII

Border Route and Branches

SHREWSBURY AS A RAILWAY CENTRE

Squeezed between a Norman castle and Telford's prison and partly built over the river Severn, Shrewsbury's Tudor style station, is a sad sort of place, far too big for modern railway needs. Yet for years it was inadequate. As J. T. Lawrence noted in the 'Notable Railway Stations' series in the *Railway Magazine* in December 1905, it had through coach services to more than thirty destinations from Penzance to Aberdeen.

> These are terminal points. The actual services are enormous. Penzance, for instance, means Exeter, Plymouth and practically all Cornwall; and London means Wolverhampton, Birmingham, Leamington and Oxford. And the bulk of these through connections only came into existence—and, in fact, were only possible—after the opening of the Severn Tunnel.

Shrewsbury was tremendously busy: 'those on the spot claim that more traffic is interchanged and redistributed at Shrewsbury than even at York.'

Despite Shrewsbury's name being featured in the title of several important early railways, they were built not so much to serve it as to join up with each other to create trunk routes. The town itself was never big. Its population had only reached 26,000 by 1881 and today it is only a little over twice that size. The Shrewsbury & Chester Railway (Vol 11) and the Shrewsbury & Birmingham (Vol 7)—'the fighting Shrewsburys' as they were dubbed from the conflict surrounding their birth—were completed in 1848 and 1854. In between came the Shrewsbury & Hereford. Henry Robertson,

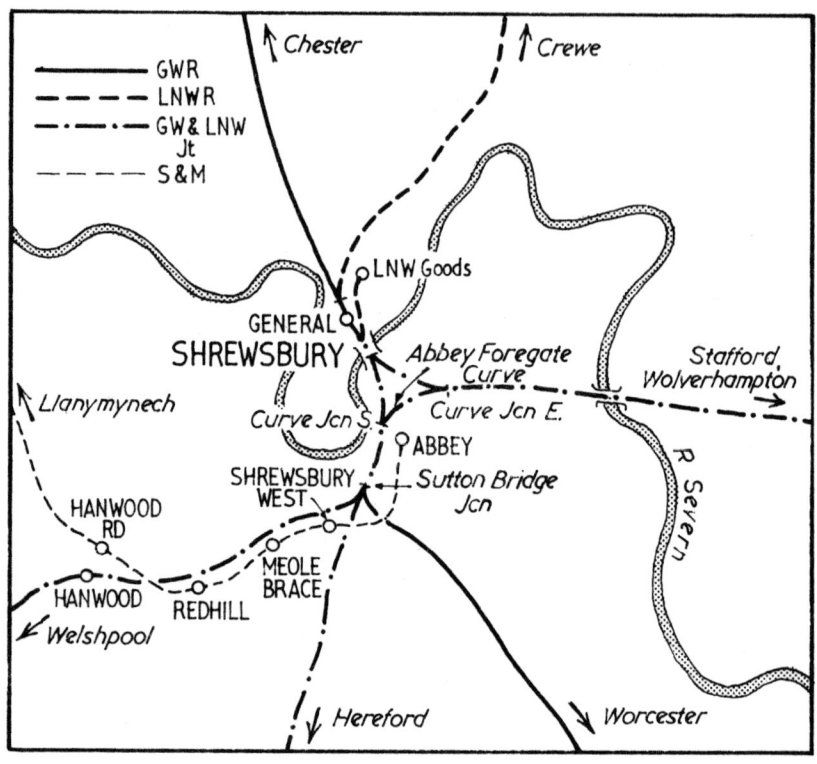

three times Liberal MP for Shrewsbury between 1862 and 1886, was engineer of all three companies.

The original station, opened in 1848, was a joint venture of the companies together with the Shropshire Union Railways & Canal Company, which with the GWR jointly owned the Birmingham route as far as Wellington, junction of the SUR's Stafford branch of 1849, the only link between Shrewsbury and London until the Birmingham line was completed in 1854. A year earlier, the Shrewsbury & Crewe Railway (Vol 7) had been authorised. Opened as a single line in 1858, it was doubled in 1862 to cope with rapidly increasing traffic as Shrewsbury became a focal point for Mid, Central and, to a lesser extent, South Wales.

Shrewsbury got another junction with completion in 1862 of the Shrewsbury & Welshpool LNWR/GWR Joint line (Vol 11), although that suffered partial traffic loss when the Cambrian

Railways completed its Whitchurch–Aberystwyth main line via Oswestry and Welshpool in 1864. Meanwhile there were two developments which produced far more traffic for Shrewsbury; the LNWR's lease of the Merthyr Tredegar & Abergavenny Railway, and Euston's drive into Mid Wales through purchase of the uncompleted Knighton Railway, stemming from the S & H at Craven Arms. When the route was completed in 1866–7, rail mileage between Shrewsbury and Swansea was reduced to ninety-five miles,—fifty-five fewer than via Newport, a route Euston had been using to compete against the Midland for South Wales after it reached Swansea via Hereford and Talyllyn.

To cope with extra traffic, the LNWR and GWR opened a quarter-mile avoiding line between the Hereford and Birmingham routes—the Abbey Curve—on 1 May 1867. Shrewsbury's final traffic boost, and one of its most important, came when the LNWR and GWR introduced the 'North-to-West' express route from the moment the Severn Tunnel opened in 1886. It was the GWR's first cross-country service and it was an immediate success, not least because it meant many long-distance travellers no longer had to change on to the Midland at Birmingham and off it again at Bristol.

But there was a price to pay: Shrewsbury station, enlarged in 1855, now had to be virtually rebuilt in 1901. The prison wall had to be underpinned and the Astronomer Royal was called to make vibration tests to make sure that it would not fall down. One of the main features of the work was digging out the station square to a depth of some 12ft so that the booking and other offices would be under the platforms, which were extended over the river.

The 'Joint' nature of this grand junction of the Northern Marches protected it from major change at Grouping. It lay within the GWR Chester Division, whose 1925–5 Report referred to arrangements with the LMS being of a 'very intricate character.' The station was controlled by a Joint Superintendent, and his District Inspector (Class 2) supervised lines to Oxley, Crewe, Madeley (Salop), Wellington, Harton Road and Buildwas. Shrewsbury's freight yards were busy: Coton Hill on the Chester line could hold 582 wagons, though in 1924 it was handling a daily average of 1,000, which arrived and departed on seventy-five trains. A private company, Hall Lewis, had a wooden wagon repair 'hut' and a siding in the yard and repaired forty-five wagons a week. In the main yard, more than twenty shunters and four locomotives worked a weekly schedule of 416 booked shunting hours, ranging over eighteen Up sidings (including Gas, No 12, and Dawson City, No 13) and five Down sidings. They handled trains to and from Oxley, Wellington, Market Drayton, Pontypool Road, Cardiff, Croes Newydd (Wrexham), Gobowen, Saltney, Chester, Manchester and Birkenhead. In the Down Goods Loop just outside the Yard, a pumping engine serviced water columns and the station's hydraulic lifts.

Castle Foregate yard (later Coton Up) handled mainly minerals, cattle, timber (some 500 tons in 1924–5) and general goods. After Tuesday auctions, an average of sixty wagons was dispatched, chiefly to stations towards Birmingham.

At another yard, Coleham, the companies exchanged words as well as traffic, the GWR complaining:

> The LMS break up a considerable number of their own trains at Coleham and stable their traffic there for some length of time. This is contrary to the spirit of the user of the Joint sidings and recently attention has been directed to similar arrangements being made at Abbey Foregate (Shropshire Sidings). It is desirable that the Great Western Company should keep a close watch on this development.

Even after Nationalisation, Shrewsbury's freight yard retained a pre-Grouping flavour, Western Region Service Time Tables of 1951 detailing eleven pilot services. They were to be worked by LNWR 0–8–0s and 0–6–0 coal locomotives ('Cauliflowers'); a Midland 0–6–0 was to have charge of another working. Other local services were to be handled by Class 5 mixed-traffic 4–6–0 locomotives.

Shrewsbury presented locomotive crews with problems. 'Special Instruction to Drivers—Blowing off Steam, Crewe Junction,' current in 1933, warned them of complaints by passengers using the station refreshment rooms. Enginemen were also warned to take special care to avoid water being splashed on to the platform if they had to use the column opposite the ladies' waiting room.

Plate 21
Above: Across the hills from Much Wenlock lay the Cleobury Mortimer & Ditton Priors Light Railway. A 1920 mixed train taking water at Ditton Priors. (*Ken Nunn collection, courtesy Locomotive Club of Great Britain*)

Plate 22
Below: Cross country – 1904. East & West Junction Railway tank No 1 dwarfed by one of the latest Great Central slip coaches. Stratford-on-Avon, 23 May 1904. (*Ken Nunn collection, courtesy Locomotive Club of Great Britain*)

Plate 23
Above: Presteign station in the early years of the present century with an admiring crowd of schoolboys looking on while the station staff and crew of Dean 0-4-2T No 1475 pose for the photographer. (*Mair/Lens of Sutton*)

Plate 24
Below: A Stephenson Locomotive Society special run to celebrate the Leominster–Kington centenary was among the best patronised trains to run on the branch. (*Lens of Sutton*)

YEARS OF DECLINE

Shrewsbury felt snubbed when it lost its London through services after withdrawal of Paddington–Birkenhead expresses in March 1967 on completion of Euston electrification. But after a fight, taken to Parliament, it has got back several daily expresses extended from Wolverhampton. This is Shrewsbury's only Inter-City route, those to Chester, Crewe and Cardiff being secondary. The Border route, despite having lost 'North-to-West' expresses when they were diverted via Birmingham in 1970 to take advantage of electrified lines north of the City and to serve larger towns than those on the route, has not been run down to the extent once feared, and is used by some summer holiday expresses to and from the West of England.

Shrewsbury's passenger network received a boost when the former Cambrian line between Whitchurch and Welshpool closed in January 1965, leaving Shrewsbury as the only gateway to Aberystwyth and the Cambrian coast, since lines from Carmarthen, Ruabon and Bangor closed at about the same time. Shrewsbury has lightly escaped passenger economies as the only ones have been those to Minsterley in 1951, through the Severn Valley in 1962, and to Stafford the following year, after the withdrawal of Wellington–Stafford local trains.

Shrewsbury is operationally important for preserved steam, introduced along the Border in 1972 and extended to Chester in 1975, locomotives taking water from a pipe at the town's new fire station, conveniently alongside the Chester line.

Freight decline has been marked and was underlined when BR abandoned plans for a major marshalling yard at Walcot on the Wellington line to replace local ones and handle then-growing inter-regional traffic. Like Brockthorpe, near Gloucester, it had been included in the 1960 British Transport Bill. That was the year in which Coleham Yard closed, to be followed by the smaller Shropshire Union (LMS) Yard in Howard Street, reached by a sharply-curved line through the station wall beside the prison. The yard had been built on the site of the SU Canal terminus.

SHREWSBURY & HEREFORD JOINT LINE

From the South Shropshire hills you can still enjoy watching steam trains go by, just as I did more than thirty years ago, their approach to Ludlow betrayed by long trails of white smoke drifting across the green fields lying to the north. Steam works hard on this route, and by leaning out of the window of a double-headed North-to-West express pounding up the stiff climb from Shrewsbury into the hills around Church Stretton you got a far better impression of the terrain that navvies had to conquer than was ever possible by examining a gradient profile.

The S & H promoters were men whose initiative built the Shrewsbury & Chester; they took time to get through Shropshire, although the route lacked the need for viaducts like those near Ruabon. The S & H was incorporated as a standard gauge line after Brunel failed to get a broad gauge foothold in the Border. Four years elapsed before construction began in 1850 when Brassey accepted not only the contract, but offered to work the fifty-mile route at his own risk, paying $3\frac{1}{2}$ per cent of the cost, and beginning an eight-year lease in summer 1854. Six years later the line was doing so well that the S & H paid six per cent dividends on its income from Brassey.

Until Nationalisation the S & H was a Joint Line as a direct result of the creation of the West Midland Railway in 1860. By opposing its formation, the S & H was given running powers over the NA & H, conceding similar ones between Hereford and Shelwick Junction for the WMR's South Wales traffic via Worcester. The Lease agreement between the WMR and the GWR alarmed the Shrewsbury directors because it gave Paddington control of its outlets at both ends. They were worried already by the impending completion of the Worcester & Hereford and Severn Valley branch, both of which were likely to drain away traffic. They approached the LNWR, which quickly offered a perpetual lease, suggesting that the GWR should join the arrangement.

Paddington reacted angrily, fearing that with its powers over the S & H, Euston would reach South Wales. The rest of the story is equally complex. The Shrewsbury and Euston directors proceeded with a Bill, but only for an LNWR lease. The GWR and West Midland naturally opposed the move, but after they were defeated in Committee, both agreed to be parties to leasing the Shrewsbury line from summer 1862.

By then the line was ten years old, having opened between Shrewsbury and Ludlow (27½ miles) on 21 April 1852 and through to Hereford for goods from 10 July, passenger services not starting until 6 December 1853. Shrewsbury–Ludlow was double track, but the section to Hereford was not doubled until a second bore was added in 1893 to Dinmore Tunnel, which burrowed 1,056yd under a wooded ridge a few miles north of Hereford.

The 'link-in-the-chain' nature of the S & H was reflected by its use of a joint station at both ends. At Hereford, it shared Barr's Court station with the GWR-worked 22½-mile single Hereford Ross & Gloucester when it arrived in 1855. Its trains ran to the south end, the Shrewsbury services using the north end. The North-to-West expresses gave the Border route prestige. The *Railway Magazine* noted that 'Ludlow, Leominster Hereford and a host of sleepy old-world towns suddenly found themselves on an important main line.'

It is an importance that is slowly returning, the line having been given extra width clearances for Euro-container trains, work which involved the demolition in 1979 of a road over-bridge near Shelwick Junction and its replacement. Local trains call at four surviving stations. Most northerly is Church Stretton, where bus-stop shelters have replaced buildings of much dignity with which the railway made its own contribution to the distinctive architecture of this pleasant little spa town. At Craven Arms, Ludlow and Leominster passengers are given train information over loudspeakers by signal-men in local boxes—that for Ludlow is three miles north at Brom-field.

BRANCHES AND MINOR INDEPENDENT RAILWAYS

The SHROPSHIRE & MONTGOMERYSHIRE RAILWAY (Volume 11) retains a toe-hold at Shrewsbury, the truncated end of the Severn Valley line providing access for petrol tank wagons to Abbey Yard, the old S & M terminus, to which leave trains used to run during World War II for troops manning a huge ammunition dump in the countryside outside the town. Eric Tonks' history of the Company (Industrial Railway Society) says all that ever can reasonably be said about the line.

Another paperback history of similar calibre is that by E. S. Griffith about THE BISHOP'S CASTLE RAILWAY. He even mentions a pub that brews its own beer! The private line was authorised in 1861 from the S & H at Stretford Bridge, north of Craven Arms, (to which its passenger trains ran fitfully over the years) to the Oswestry & Newtown (later Cambrian Railways) near Montgomery. Bishop's Castle, reached 24 October 1865, was as far as it ever got, and even that journey involved a reversal amid fields at Lydham Heath. The BCR's fate was sealed when it was excluded from Grouping, and after local authorities tried unsuccessfully to persuade the GWR to buy the line in the early 1930s and extend it to Montgomery. Advocates of a new short cut to the Welsh coast ignored the point that most holiday trains had to run via Shrewsbury and direct routes to and from the West Midlands and other population centres. Since 20 April 1935, the BCR has lived only in the memory of enthusiasts who revere eccentric lines.

Under Wenlock Edge ran the CRAVEN ARMS–BUILDWAS branch, used by passenger trains to Wellington (twenty-eight miles in 1½ hours). The Much Wenlock & Severn Junction Railway of 1859 built the 3¾ miles from the Severn Valley branch at Buildwas to Much Wenlock and both opened the same day, 1 February 1862. Coalbrookdale–Presthope (including the Severn bridge still used by mgr trains to Buildwas Power Station) was completed 1 November 1864 by the Wenlock Railway (incorporated as the Much Wenlock Craven Arms & Coalbrookdale Railway in 1861). The Wenlock Railway finished this minor cross-country route when it reached the S & H at Marsh Farm Junction, north of Craven Arms, on 16 December 1867. The eleven miles had taken three years to construct. Closure was piecemeal. Craven Arms–Much Wenlock closed to passengers at the end of 1951 but ten of the seventeen miles (between Much Wenlock and Longville) remained for parcel trains—and unofficial passengers. The more industrialised Much Wenlock–Buildwas–Wellington stretch retained a passenger service until summer 1962, total closure between Longville and Buildwas being accomplished in December 1963. For years the branch had carried a variety of traffic: lime, cement, stone, timber and horses and cattle, for which most of the small, oil-lit stations had their own docks. Traffic was profitable: Longville's revenue in 1924 was £3,276. Total at Buildwas (population 300) was only £100 less.

The BRIDGNORTH CLIFF RAILWAY was not Shropshire's only cable-worked railway. For years there was the purely mineral LUDLOW & CLEE HILL RAILWAY. It included a 1¼-mile rope-worked incline, which lifted stone wagons by 600ft, mainly at 1 in 12, to a mile-long 'level', 1,250ft above sea level. This top section was worked by Sentinel or other small locomotives. Opened privately 24 August 1864, the line ran six miles from Ludlow and was tricky to operate. The approach to Bitterley Yard, at the incline foot, was at 1 in 20, and on wet days, pannier tanks could cope only with half their usual load of twenty empty wagons. The line prospered for years. After World War I, about 6,000 tons of roadstone a week left the hillside quarries by rail, but lorries gradually took over the traffic after World War II and the incline closed in November 1960, the remainder in December 1962. Like the S & H, it was a joint line for most ot its life, the owners signing a working agreement with the GWR and LNWR in 1867. It was absorbed jointly by the companies from 1 January 1893.

Another line which became joint was the 5¼-mile WOOFFERTON–TENBURY WELLS, built 1859-61 by the Tenbury Railway, with the help of the S & H, which provided land by disposing of part of the Leominster Canal. After the S & H Joint Line was created, the branch which had opened 1 August 1861, was worked as part of it by the GWR, which also worked the fourteen-mile TENBURY & BEWDLEY RAILWAY, incorporated 1860, opened 13 August 1864, and vested in the GWR five years later.

Both lines were very much rural, and when it came to economies the first to be opened was the first to go. Passenger trains were withdrawn between Woofferton and Tenbury Wells and the line closed completely in July 1961, together with Woofferton station. That prompted a *Railway Magazine* reader to recall a visit in the 1930s when the station name was spelt correctly on the nameboards, while the signal box sign read 'Wofferton' and platform trolleys stated 'Wooferton.' It had been planned to withdraw Woofferton–Bewdley passenger tains, but because of local protest, a service was retained five days a week between Tenbury Wells and Bewdley for another year: out in the morning, home in the evening. The next economy was complete closure Tenbury Wells–Cleobury Mortimer in January 1964. The remaining six miles survived until April 1965. Dismantling included the Dowles Viaduct across the Severn, a short distance from where the branch left the Severn Valley branch. The abutments and two stone piers can still be spotted from the preserved line.

It had been retained to provide access to the CLEOBURY MORTIMER & DITTON PRIORS LIGHT RAILWAY, which changed hands four times in its short career. It was built by a private company, under a Light Railway Order of 23 March 1901, absorbed by the GWR on 1 January 1922, passed to BR in 1948, was taken-over by the Admiralty from 1 May 1957 and worked by it from 30 September. The 12¾-mile line took seven years to construct, and opened to goods 19 July 1908; passengers 20 November. Part of its traffic came down a rope-worked incline from Clee Hill to the Rea Valley. Speculative extensions to the Craven Arms–Buildwas branch at Presthope and the Severn Valley branch at either Bridgnorth or Coalport were considered and soon forgotten and the Light Railway itself lost its sparse passenger service in September 1938, earlier than many Shropshire branches. But the line was far from finished, and until complete closure on 16 April 1965 it served a huge armaments depot, which the Admiralty hid in the sparsely-populated wooded countryside.

For many years a journey between Worcester and the Welsh border was tediously possible by changing at Leominster. The forty-eight miles between Worcester and New Radnor took almost three hours, including a forty-minute wait at Leominster. Kington trains reached Leominster from a junction on the S & H just north of the station. Those to Worcester departed along a single line which kept company with the main line for almost a mile before turning east into wooded hills. The WORCESTER BROMYARD & LEOMINSTER RAILWAY of 1861 with Sir Charles Hastings, founder of the British Medical Association, as chairman, was projected 24½ miles to the Worcester & Hereford at Bransford Road (later Leominster) Junction, but in 1869 the promoters abandoned Leominster–Bromyard (twelve miles). Soon afterwards the GWR agreed to work the rest and it opened to a temporary terminus at Yearsett on 2 May 1874, the remaining 3¾ miles to Bromyard following 22 October 1877. Meanwhile, the abandoned section had been revived, the LEOMINSTER & BROMYARD RAILWAY of 1874, reaching Steens Bridge ten years later, 1 March 1884, but taking another thirteen years to complete the single line through to

Bromyard on 1 September 1897. By then, both companies had been part of the GWR for nine years.

Leominster–Bromyard trains ran virtually empty for years. Shortly before total closure in 1952—twelve years ahead of Bromyard–Worcester—there were just seven season ticket holders, including a fifteen-year-old girl soon to leave school, and two daughters of a ganger, who would move if the branch closed. Afterwards the line became a cripple wagon store, batches of sixty being common with rails removed in between to avoid runaways on steep gradients. No such problems face the 2ft 0in gauge Bromyard & Linton Light Railway, using about a mile of trackbed.

THE KINGTON BRANCHES

'Officially known as the KINGTON Branch,' wrote John D. Hewitt in the *Railway Magazine* in that momentous month of September 1939, 'the trident-shaped system of lines which stretches from Leominster, in north-west Herefordshire, westwards towards the small market town of Kington and forks with Eardisley, New Radnor and Presteign as its prong points, has on the map a quaint air of detachment—almost as if it served a tiny and independent state.'

It was a 'state' where railways arrived early, the Kington Tramway of 1820 linking limestone quarries with the Hay Railway of 1818 by an end-on connection at Eardisley. The 13¼-mile LEOMINSTER & KINGTON RAILWAY of 10 July 1854, was opened to Pembridge (eight miles) in January 1856 and to Kington 2 August 1857, being leased to the GWR and West Midland in 1862. Kington was an important railhead, Murray's *Handbook* of 1860 stating that the town (population 3,200) 'is a favourite starting place for tourists to Aberystwyth, whither a coach runs daily, conveying passengers who are brought to Kington by the railway from Leominster.' Despite the comparative isolation of the countryside, the L & KR was successful enough to pay four per cent dividends, encouraging other promoters to project the KINGTON & EARDISLEY RAILWAY in 1862, although twelve years elapsed before the seven-mile line was opened on 3 August 1874, having been partly adapted from the Kington Tramway. The junction with the L & KR was at Titley, nearly two miles east of Kington, running powers being exercised by the K & ER which on 25 September 1875 carried the Kington branch 6½ miles west to New Radnor, where the terminus was unsatisfactorily sited at the foot of a hill half a mile from the village. The GWR which worked the local lines, absorbed the K & ER in 1897 and the Leominster Company a year later, including its Titley–PRESTEIGN branch of 5¾ miles, authorised in 1871 and opened 10 September 1875.

Closure was fragmented and protracted. The Titley–Eardisley branch was so rurally remote that it suffered economies in both world wars, being closed temporarily 1917–22 and permanently from 1 July 1940. Coal shortages led to Kington–New Radnor and Presteign branch passenger trains being hurriedly withdrawn from 5 February 1951, an economy confirmed in June. At the end of the year Dolyhir–New Radnor closed completely. Leominster–Kington passenger services ran into increasing bus competition and after losing a fares price battle they were withdrawn on 7 February 1955, a collection for the crew of the last train being taken in a ninety-year-old policeman's helmet. Kington–Dolyhir closed completely on 9 September 1958, Leominster (Kington Junction)–Kington and the Presteign branch following 28th September 1964.

GOLDEN VALLEY RAILWAY

It may seem silly to regard the Golden Valley Railway as a through route to South Wales, but that was the way it saw itself in a share-raising prospectus of 1888. It was to be an important link in a route between Liverpool and Bristol via the Mid Wales Railway to Three Cocks, linked in, south of Pontrilas by an extension to Monmouth to reach the lines to Chepstow and Bristol via the Severn railway bridge.

This was twelve years after the Company had been incorporated and seven since the first 10½ miles between Pontrilas and Dorstone had opened on 1 September 1881. Dorstone–Hay (8¼ miles) did not open until 27 May 1889. The line closed in 1897–98 because of financial difficulties, but was acquired by the GWR in 1899, re-opened in 1901 and proceeded to live a quiet existence, disturbed by the creation of a large storage depot (with its own locomotives) near Pontrilas during World War II.

Economies and closure, begun with the withdrawal of the sparse local passenger service in 1941, continued until 1957. The late Professor C. L. Mowat wrote a splendidly entertaining and detailed history, sub-titled 'Railway enterprise on the Welsh Border in late Victorian Times.' and other historians were attracted by its charm. John D. Hewitt noted in the *Railway Magazine* in 1938: 'Thank heavens we still have railways with character.' He also spoke of its 'drowsy mixed trains.'

SOUTH WALES SPRINGBOARD

A line which had greater, though limited, success in becoming part of a South Wales through route was one from which the Golden Valley Railway stemmed: the HEREFORD HAY & BRECON. In *Bradshaw's* 1908 Timetable, the last table in the sixty-two pages of Midland Railway services was that of Swansea, Neath, Brecon, Hay and Hereford. A Swansea–Hereford journey demanded passenger stamina, for the $79\frac{1}{4}$ miles with twenty-four intermediate stations took four hours, including a fifteen-minute reversal at Brecon. Only seven stations were on the HH & B, which had generally few passengers, one reason why the line closed as Dr Beeching was preparing his Report. I was on the valedictory train, run by the Stephenson Locomotive Society on 30 December 1962, which first wiped off the map, the Mid Wales Railway. By the time it left Brecon for Hereford, en route to Shrewsbury, it was dark and my only memory of the HH & B is of snow-covered fields seen from the carriage window. Incorporated 8 August 1859, the 34-mile HH & B was truncated before construction when only a year later, Talyllyn–Three Cocks Junction was transferred to the Mid Wales Railway, and Talyllyn–Brecon to the Brecon & Merthyr. The bid by different companies to get to Brecon developed into a major battle (see *The Cambrian Railways* Vol 1), and the Hereford company bought the Hay Railway in 1860, adapting three miles, half between Pontvain and Clifford Castle (see C. R. Clinker: *The Hay Railway*, 1960). The HH & B was built up the Wye Valley in stages: Hereford –Moorhampton (nine miles) opened to goods on 24 October 1862; Moorhampton—Eardisley (five miles) on 30 June 1863; Eardisley– Hay (seven miles) on 11 July 1864. Hay to Three Cocks Junction

(5½-miles) followed on 19 September 1864, when through services to Brecon began. Savin, as contractor, worked the HH & B. Among early passengers was the Rev. Francis Kilvert, who recorded rail journeys in his famous diaries. (See A. L. Le Quesne: *After Kilvert*, 1978).

Amalgamation was carried out with the Brecon & Merthyr, but it was ruled illegal. The Midland Railway, which had started running goods trains to Hereford via Worcester in November 1868 and passenger trains the following July, saw its opportunity of getting into South Wales. It leased the HH & B in the same year and progressively increased its hold, absorbing the company in 1876. By then, relations with other companies had settled down, though not before a classic engine and wagon blockade by the GWR in 1869 to keep the Midland trains out of Barton station at Hereford. The HH & B was forced to adapt a temporary terminus at Moorfields station and use it for five years. Almost two decades later passenger services settled into the pattern they followed until withdrawal, reaching Barrs Court station, Hereford, over a short link to Barton & Brecon Curve Junction opened in 1893 by the Shrewsbury & Hereford Joint (page 111). The Midland never attempted ambitious developments between Hereford and Swansea, being content with only local passenger services. The route's value lay in through freight traffic, including a busy link which the LNWR developed between South Wales and Birmingham over the HH & B. Through traffic survived into the 1960s, the HH & B remaining open until cut back from Talyllyn Junction to Eardisley on 4 May 1964 and to Hereford (Brecon Junction) four months later. A well-visited stretch survives at Hereford as part of Bulmers Railway Centre.

CHAPTER VIII

The Forest of Dean and Wye Valley

The unusual and especial flavour and character of the one-time maze of Forest of Dean lines is enshrined by the Dean Forest Railway Society Limited (entry No 356 in *Steam '80* handbook of preserved lines) recorded with precision in H. W. Paar's two volume history *The Severn & Wye Railway* and *The GWR in the Forest of Dean*, and beautifully reflected by photographer B. J. Ashworth who sees in railways more than trains. His Forest of Dean section in *Steam in the West Midlands & Wales* is of freight lines which were never as well known as the railways through the neighbouring Wye Valley. One reason is that while the Wye lines remained open to passengers until the 1950s, one of the main passenger routes through the Forest was closed as early as 1929 through the withdrawal of passenger services north of Lydney Town.

WEALTH OF COAL AND IRON

The Forest, an area of some thirty-six square miles, was rich in minerals from Roman times. In medieval days, it was Britain's main iron-working centre; it enjoyed a boom in Victorian times and earlier because its huge coal deposits were shallow-lying and could be mined in eras before it was possible to reach deep-lying seams in South Wales and elsewhere. Industry attracted a large influx of workers, and the Forest of Dean Parliamentary Division of Gloucestershire of a century ago had 53,000 voters—17,000 more than Gloucester.

Two distinct, though connected, systems of railways developed—

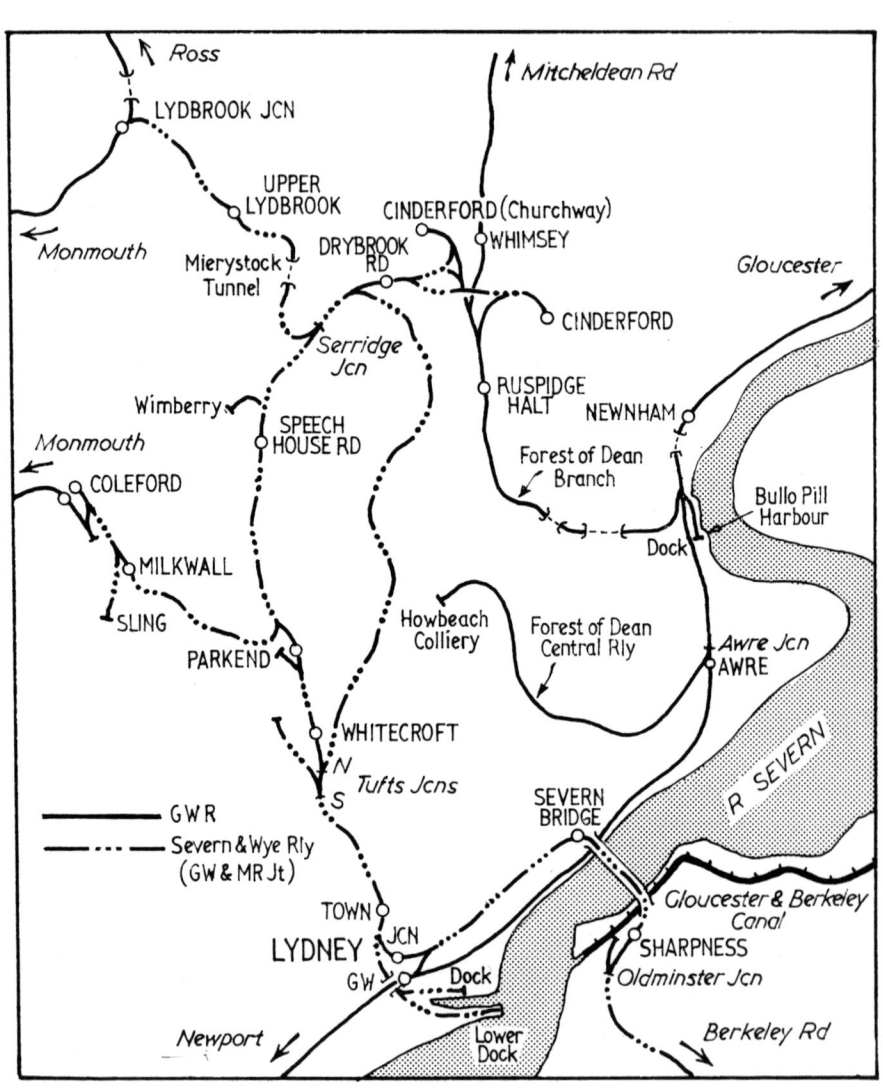

Ross

LYDBROOK JCN

UPPER
LYDBROOK

CINDERFORD (Churchway)

Mitcheldean Rd

WHIMSEY

DRYBROOK
RD

Mierystock
Tunnel

Monmouth

CINDERFORD

Gloucester

Serridge
Jcn

Wimberry

SPEECH
HOUSE RD

RUSPIDGE
HALT

NEWNHAM

Monmouth

COLEFORD

Forest of Dean
Branch

Bullo Pill
Harbour

MILKWALL

Dock

SLING

PARKEND

Howbeach
Colliery

Forest of Dean
Central Rly

Awre Jcn
AWRE

WHITECROFT

R SEVERN

N
S

Tufts Jcns

SEVERN
BRIDGE

———— GWR

————— Severn & Wye Rly
(GW & MR Jt)

TOWN

JCN

LYDNEY

GW

Dock

Gloucester & Berkeley
Canal

SHARPNESS

Oldminster Jcn

Newport

Lower
Dock

Berkeley Rd

the Severn & Wye Joint lines of the GWR and Midland (later LMS), surrounded by purely GWR routes: Gloucester–Newport section of the South Wales Railway; Hereford Ross & Gloucester secondary route; the Monmouth–Coleford branch; the Forest of Dean Central Railway; the FOD branch and the Mitcheldean Road & Forest of Dean Junction Railway.

EARLY TRAMROADS AND SEVERN & WYE RAILWAY

Forerunners were horse tramways, three major ones being completed within six years from 1809. One was developed by the Severn & Wye Railway & Canal Company, connecting both rivers and broadly serving the western Forest. Formed as the Lydney & Lidbrook (*sic*) Railway Company in 1809, it became the Severn & Wye Railway & Canal Company a year later, the canal element being a one-mile stretch from the Severn to Lydney. It opened in 1813, a year after the main tramroad (laid on stone blocks) had been completed, together with eight short branches. While impressive in length (it grew to almost thirty miles), the system was primitive. More branches followed, but little development took place until completion of the South Wales Railway.

Early relations between the SWR and the Severn & Wye were uneasy and complicated ,the S & WR resisting attempts to take the broad gauge into the Forest. But at the suggestion of a House of Lords Committee, the SWR agreed to contribute £15,000 towards improving the S & WR.

That was in 1847. Four years later the South Wales Railway was opened from Grange Court to Chepstow and soon traffic was being exchanged at Lydney.

It was not until 1853 that the S & WR was authorised to supply motive power and become carriers. Even though the main line was broad gauge, the same S & W Act authorised further tramways of $3\frac{3}{4}$ miles, with 'Gauge of $3\frac{1}{2}$ feet.' The Company stated in Bradshaw's *Manual*:

> The present line for conveying produce of Forest of Dean to Lydney Harbour is a descending line throughout used exclusively for minerals; gradients such as to give impetus to loaded carriages, and to take back empty carriages, 1 in 40 steepest.

It was a decade later before steam locomotives were tested, very much experimentally, on such gradients. But they proved an immediate success and changed the whole character of the lines in the Forest.

Stiff gradients remained after the transition from tramroad began with seven miles of broad gauge line (because the SWR was paying) from Lydney up a wooded valley to Wimberry (Speech House Road) on 19 April 1869. Re-gauging to standard took place in May 1872 and a major extension of the main line followed on 26 August 1874 with a 4½-mile line from Serridge Junction (north of Wimberry) to Lydbrook Junction on the newly-opened Ross & Monmouth Railway. It forged a direct link between the Forest's ore mines and the ironworks of South Wales.

The main line climbed all the way to Serridge on gradients which stiffened to 1 in 40. From there, the line ran downhill for three miles at 1 in 50, but the stretch is perhaps best remembered for Lydbrook Viaduct, depicted by an oil painting on the dust jacket of *The Severn & Wye Railway*—an idyllic late Victorian scene. The view from the village was different, for until 1965 the viaduct spanned the main street at a height of 90ft.

The main line assumed its final shape on 2 July 1900 with an extension of just under a mile from Cinderford (Old) station to one more central. This was six years after the S & WR had been rescued from financial difficulties and taken-over by the Great Western and Midland. The Cinderford station was used jointly from 1908 after the GWR built the half-mile Bilson Loop for trains running from Gloucester over the Forest of Dean Railway.

MINERAL LOOP AND LYDNEY HARBOUR LINES

The main line between Wimberry and Drybrook Road, on which Serridge Junction was situated, was regarded technically as part of the Mineral Loop, the other main part of the S & WR. The 6½-mile Loop was constructed after victory in a Parliamentary battle against the GWR and Forest of Dean Railway, in which the S & WR convinced MPs that the Loop would provide the best means of getting coal from several highly productive pits to the sea. The first half-mile from the main line at Tufts Junction, north of

Lydney, to Pillowell was laid broad gauge, but the Loop was completed standard gauge on 22 April 1872.

Heavy gradients made the Loop difficult to work and after several accidents caused by irregular working, it was split into three operating sections with rurally-named boundaries: Drybrook Road–Crump Meadow Colliery Sidings; from there to Foxes Bridge Colliery Sidings and south to Tufts Junction. Train staffs were kept in padlocked boxes fixed to signal posts and the keys, which also opened padlocks securing sidings points, were held by guards.

The harbour remained the main outlet for the S & WR coal traffic despite connections with the GWR at Cinderford and Lydney. The docks branch (opened 9 December 1875) crossed the South Wales main line on the level at Lydney Junction and split into two arms: the longer ran 1¾ miles to the Lower Dock, the shorter stretched just under half-a-mile to the Upper Dock.

As a port, Lydney faced some competition from Bullo Pill, also with Berkeley and Sharpness across the Severn. Apart from a restricted entrance width, which proved an increasing handicap as ships got bigger, there were also difficulties in maintaining the water levels. In its heyday, up to 300 small sailing ships crammed into the port waiting to take coal to small ports in Devon like Ilfracombe, to the Scilly Isles and, nearest of all, Bristol.

COLEFORD AND OTHER BRANCHES

Besides building its own outlets from the Forest, the S & WR had to deprive the GWR of access with a line from the east which would rival its own ore route to South Wales. The GWR's threat was to iron-ore around Coleford, rather than Cinderford. Coleford, capital of the Forest with a population around 3,000 people, attracted tourists as well as industry, *Baddeley* stating 'which considering it is a mining centre, is by no means an unsightly little town. The situation is picturesque, in broken country 500ft above the sea.'

The threat of GWR existed for years, for Coleford had been promised a railway for more than two decades before the S & WR arrived—ever since the creation in 1853 of the Coleford Monmouth Usk & Pontypool Road (page 130). The S & WR's four-mile Coleford branch from Parkend was authorised in 1872 and opened for goods on 19 July 1875, passenger services following on 9 December, some weeks after their introduction on the main line.

Once again gradients were steep, just within steam locomotive capability. Two-and-a-half miles of the branch were at 1 in 30, little wonder that Service Time Tables (*sic*) Appendix laying down that 'No Wagon with a defective brake must, under any circumstances, be allowed to work on the Branch.' The branch's value was not solely in keeping the GWR away from the area; it was the stem from which short lines, again steeply graded, were built to serve quarries and iron-ore mines. Best-known was perhaps the Sling Branch, half-a-mile at 1 in 40 to an ore mine.

A number of other branches, (including the colliery-serving Oakwood of similar length, which formed the third constituent of Tufts Junction), are shown on a Joint Lines Railway Mileage Map reprint of 1923 (with notes by C. R. Clinker), published by Avon-Anglia.

SEVERN BRIDGE RAILWAY AND SHARPNESS BRANCH

Dismantling of the Severn railway bridge was completed in 1970, four years after the Queen opened the suspension bridge further downstream, the first road bridge across the Estuary. Yet it might not have been, for among proposals considered when the railway bridge was projected a century earlier was one for it to carry a road as well as the line. The road idea was ahead of need, but the railway bridge of 1,387yd was a little masterpiece, which might have attracted more attention but for being constantly overshadowed by the Severn Tunnel. But unlike the tunnel, opened only seven years later, the bridge was inadequate, being conceived to branch or at any rate secondary line standards rather than those of a main line. Rivalry between companies was partially responsible, for had the Midland got access to the Forest system over the GWR from Gloucester it is unlikely to have contributed towards the bridge since it would have given it few advantages in meeting traffic competition.

As it was, the bridge was built by the independent Severn Bridge Railway Company of 1872, financially supported by the Midland and the S & WR. It was to have a line from the GWR main line at

Plate 25

Above: Pontypool Road, which looked – and was – one of the most important junctions on the Welsh Border. Imposing station to match, now demolished. (*G. H. Platt*)

Plate 26

Below: Watching preserved steam go by is pleasant and exciting. *King George V* passes the site of Marsh Brook station with a Wirral Railway Circle special on 5 October 1974. (*R. W. Miller*)

Plate 27
Above: Coaley – junction of the Dursley branch on which veteran tank locomotives spent their last years. The signalbox separates Jubilee class No 45626 on a Nottingham–Bristol express from the branch locomotive, No 41720 in December 1955. (*Real Photographs*)

Plate 28
Below: Ashchurch with its two junctions. The main Bristol–Birmingham line is in the centre, the branch to Tewkesbury and Malvern diverging to the left, and the loop line through Evesham and Redditch to Barnt Green to the right. (*L&GRP, courtesy David & Charles*)

Lydney (as well as the S & W system), to Sharpness, to join a purely Midland branch, authorised the same day (18 July) from the Bristol & Gloucester main line at Berkeley Road. Bridge construction, effectively from 1875, was straightforward, although not without incident. A BBC colleague's grandfather, who fell from a pier and landed on sand below, was presumed dead. He regained consciousness while being carried away in a sack for burial.

The bridge deck, 70ft above high water, was supported on twenty-one spans of bow-string girders on cylindrical columns. The Gloucester & Berkeley Canal was crossed on a 400-ton swing bridge, worked by steam; 'the gauge cocks to be blown through at least once a day,' stated the operating instructions.

The opening day of the bridge and the four new miles of railway between Sharpness and Lydney Tin Works Junction (S & WR) and Otters Pool Junction (GWR) was significant for a number of reasons. It was in the centenary year of the famous Iron Bridge, which also crossed the Severn, although miles upstream; it was a day when, as it happened, the Severn Tunnel workings were disastrously flooded, and it was only weeks ahead of the Tay Bridge disaster. It was also the day on which the S & WR and the Severn Bridge companies amalgamated—an agreement which did not have any bearing on the Sharpness branch of the Midland, which also had a short branch to the local docks, crossed on a swing bridge.

The Sharpness branch, also of four miles, opened ahead of the bridge, reaching Sharpness from Berkeley Road on 2 August 1875, being used by passengers from 1 August 1876. The docks branch opened the same day, and there was a slight change to passenger services three years later when a new station was opened at Sharpness with the bridge.

The new company was soon in financial difficulty, a situation which led to its purchase jointly by the GWR and Midland from 1 July 1894. The price was nearly £500,000 and the deal included the transfer to the Joint Line of the Sharpness branch of the Midland, which was refunded half the capital cost by the GWR. The Joint system brought a number of improvements and changes while the S & WR's handful of locomotives were shared between the big companies. The Joint system stretched for thirty-five miles and all but three miles (Otters Pool Junction–Lydney Town, and Tufts Junction–Coleford Junction) were single. The Sharpness branch was to be singled in 1931. The system might have been longer had a scheme gone ahead for the Severn & Wye, with Midland support, to build a connection from Berkeley Road via Nailsworth to the Midland & South Western Junction at Cirencester, mainly to get coal to Southampton. It was considered several times, perhaps never too seriously. Berkeley became a more important junction when the east–south Loop to the Midland main line was opened 9 March 1908 to give the GWR more direct access from the Badminton route via Yate.

The Lydney–Sharpness–Berkeley Road section was used as a diversionary route for the Severn Tunnel, goods trains from South Wales to the West of England cutting the distance between Chepstow and Bristol by thirty miles, compared with continuing via Gloucester which, incidentally, was the GWR base for relief signalmen for Joint Line boxes, and civil engineering staff trained to work the Severn swing bridge.

The bridge's fate is well known. It was hit by the 229-ton petrol tanker *Wastdale* on the foggy evening of 25 October 1960. Five lives were lost in the accident, which happened seven minutes after a train had crossed. Afterwards, Berkeley South Loop fell into disuse, being closed in January 1963, but the Sharpness branch remains valuable for traffic to Berkeley nuclear power stations, also atomic waste for shipment from Sharpness.

Speculation about what had happened if the bridge had not been severed is futile, though it is doubtful if it would still be in use. Beeching would hardly have considered local passenger services as viable, while the line could not have been used by Inter-City diversions.

THE GWR IN THE FOREST

Broadly, the GWR web of lines comprised branches serving areas east of the Severn & Wye system, notably Cinderford. The most historic was the Bullo Pill tramroad of 4ft 0in gauge, also dating from 1809. The gauge has never been certainly established. (*The GWR in the Forest of Dean*, page 29). What is known is that the South Wales Railway was authorised on 2 July 1847 to lease or buy and reconstruct the tramway and when this was accomplished on 24 July 1854 there were 7¼ miles of broad gauge from Newnham to Churchway and other pits, together with a branch more than a half-mile long, from the main line to a little harbour at Bullo Pill; a basin of under 100yd long and about a fifth of that width, difficult of shipping access, not least because of the fast tides. Easier to reach were riverside wharves served by the tramroad.

The only major work necessary for the broad gauge was widening Haie Hill tunnel, 1,064 yds, under Blakeney Hill, probably the first built on a public railway in Britain. It took place while coal traffic still ran through, for the local pits were booming. An easier engineering operation was conversion of the line to standard gauge in 1872. The following year the branch was connected to the Severn & Wye at Bilson Junction, near Cinderford and the curve to the Joint station followed in 1908 (page 122). The branch was busy for years, receiving a boost from the opening of Northern United Colliery, near Cinderford in 1935. Production ceased after thirty years.

The Forest of Dean branch was extended northwards from Bilson Junction, Cinderford, by the Mitcheldean Road & Forest of Dean Junction Railway, projected to the Hereford Ross & Gloucester Railway of 1855 (page 134) to serve and provide an outlet for mines and quarries in the north of the Forest. The Great Western and Severn & Wye companies opposed the projected line, seeing it as a threat to their traffic, but it did not prove so, since it was never opened throughout. Incorporated in 1871 to run no more than 4¼ miles, it was dogged by the need for heavy engineering works and gradients and the economic climate. It was absorbed by the GWR in 1880, but the first 1¾ miles from Bilson Junction to Speedwell Siding were not opened until July 1885. Steam railmotors were introduced on a service from Newnham when a half-mile extension to Drybrook Halt opened 4 November 1907. The remaining 2½ miles, including Euroclydon Tunnel (638yd) was laid with track, and although never opened was maintained until Drybrook Quarry–Mitcheldean Road was taken-up for scrap in February 1917.

FOREST OF DEAN CENTRAL

The broad gauge conquest which began with such magnificent main lines like the London to Bristol, faded quietly away in the Forest of Dean on a branch that had a sad existence. 'One of the most unfortunate of small railway concerns,' was how historian John Marshall described the Forest of Dean Central Railway in a valediction in the *Railway Magazine* of June 1958. It suffered a number of setbacks after being projected to link collieries in the Howbeach Valley with a small dock planned on the Severn at Brimspill, between Bullo Pill and Lydney. The harbour was never built, partly because the coal-owners rejected appeals for financial help, and the pits were better served by the Severn & Wye Mineral Loop, which was connected to a dock. The Forest of Dean Central also failed to become (as once was thought) a through route to the Hereford Ross & Gloucester, again near Mitcheldean Road with an extension to Ledbury. Incorporated in 1856, the FODCR was in immediate financial difficulties and soon admitting that 'a great number of shareholders had not paid up,' though optimism reigned. The company stated in 1859 'Since the line was projected large coalfields had been got into working order, and consequently the expectations formed with regard to traffic had been so far realised.' Had they? That was stated at a time when there was just £3.00 in the secretary's hands. It was not for almost another decade (25 May 1868) that the line opened from the South Wales main line at Awre to Holbeach and New Fancy Colliery. By then it belonged to the Great Western and the 4¾ miles were among the last laid to broad gauge. In that form they survived (with several other Forest stretches) for only four more years.

GWR TO COLEFORD

To complete the story of railways that served the Forest, rather than skirted it like the Hereford Ross & Gloucester, are two lines of no especial significance by which the GWR reached Coleford with a route by no means as direct or useful as the Severn & Wye's from Parkend, within the Forest.

The Coleford Monmouth Usk & Pontypool Road was authorised

in 1853 as a private company headed by the prominent ironmaster Crawshay Bailey MP, created on the incentive of the owners of Ebbw Vale and Dowlais ironworks after they acquired sources of ore and coal in the Forest. Apart from being shorter than lines via Newport, the line was to be more easily graded than the branches climbing the South Wales valleys. It was planned from the Newport Abergavenny & Hereford at Little Mill, about two miles north of Pontypool Road, to Coleford. A half-mile branch from Dixton to Monmouth Gas Works was never built.

The promoters saw it as something grander than its title implied: a section of a line through the Forest to Gloucester and London. It had powers to take over the first local tramroad, the well-established Monmouth Railway, completed between Monmouth and Coleford in 1817. The CMU & PR itself reached Usk (4½ miles) 2 June 1856 and Monmouth (Troy), a further thirteen miles from Little Mill Junction, on 12 October 1857, the company reporting two years later 'The passenger traffic to Monmouth already exceeds estimates submitted to Parliament for the whole line to Coleford.'

Money was short, but a viaduct was built across the Wye so that traffic could be transhipped at Wyesham Wharf from 1 July 1861. This was the year when the small company was leased by the West Midland Railway, which had far more pressing problems and did nothing about an extension to Coleford. Wyesham remained forever the eastern boundary of the line.

Hopes were raised when the Worcester Dean Forest & Monmouth Railway was authorised 21 July 1863 between Great Malvern and the CMU & PR at Dixton Newtown, via Mitcheldean and Coleford. But after financial troubles of its own, the scheme was reduced to Monmouth–Coleford by Board of Trade Warrant of 8 December 1868. Again, there was no construction done; eventually a GWR-inspired company, The Coleford Railway, was incorporated in 1872. Because of a general depression and the Severn & Wye's completion to Lydbrook Junction, there was little incentive for the Coleford Railway, and 5¼ miles were not completed until 1 September 1883.

The only access between the companies at Coleford was through a shunting neck, a situation not rectified until 26 October 1951 when a direct connection allowed through running from Parkend to Whitecliff Quarry, which had been the terminus of the Coleford branch ever since the remainder to Wyesham Junction at Monmouth had been lifted 31 December 1916, a wartime economy which became permanent.

DECADES OF DECLINE

All the Forest lines are now lumped together to show the lengthy years over which they suffered decline and broadly the pattern of it. The value of the lines that grew out of the early and primitive tramways in carrying the mineral and goods traffic remained for years, while passenger services introduced over them were sparse and shorter-lived. The railways of the Forest were created and died as the industries they served were created and died. It was as simple as that. There were no complicating factors like the birth of new industries or towns. Colliery closures stretched over almost a century. By 1937 the LMS Bristol District Goods Manager had just seven left to serve, including three on the Mineral Loop. From 1930 the pits had a centralised selling scheme, based in Coleford. Most was rail-borne: in 1950 BR was still handling about two-thirds of local coal output. The Western Region, which inherited the Severn & Wye Joint Line from the London Midland on 2 April 1950, moved 558,000 tons that year, earning £303,000 from the coal and other local traffic.

While passenger economies started before World War II, the withdrawal of goods services and line or section closures did not begin until after Nationalisation. This was partly because the railways had been busy through the Forest, which had been a huge, well-hidden and dispersed ammunition dump. The Severn & Wye main line closed in sections, a pattern accentuated when Serridge Junction–Lydbrook Junction and the Mineral Loop closed in 1956. The Lydney Dock branch followed in November 1960, although the crossing on the main line survived another three years. Lydney locomotive shed closed in 1964, but the Coleford branch handled Whitecliff Quarry traffic until 1967 when it closed together with the stretch north of Parkend. The scene was set for enthusiasts who had formed the Dean Forest Railway Preservation Society (which produces an informative guide about itself, other lines and their remains) to begin putting steam back into the Forest. They hope to acquire the 3½-miles between Lydney and Parkend, which BR closed late in 1980.

Of the former GWR lines, the Forest of Dean Central Railway was little-used for years before closure in August 1949. It was perhaps more useful for the next ten years when it was retained for wagon store. The Forest of Dean branch remained in use until August 1967, although it had lost its Bullo Pill Dock arm four years earlier.

With the passenger services, the first to go were those of the Severn & Wye, six years after Grouping, not that this had profoundly affected the system. The three services withdrawn north of Lydney Town were those to Cinderford, Coleford and Lydbrook Junction, the last being the only one that ran through the Forest rather than terminating in it. They had been introduced late in 1875 and were withdrawn on 8 July 1929. All had been lightly used, not least the Cinderford trains because there was the more direct route to Gloucester via Newnham.

The through service from Berkeley Road to Lydbrook Junction was hardly attractive because it took an inordinate length of time—about 1½ hours for what *Bradshaw* showed as twenty miles but was actually further—the trains twice ran between Serridge Junction, where reversal could have taken place, and Cinderford, where it did.

A year later, the tip of the Forest of Dean branch was closed to passengers between Cinderford and Drybrook Road, but the branch retained passenger trains until November 1958 after the failure of an experiment to meet increasing bus and car competition by extending trains from Newnham to Gloucester (Central). It was given a lengthy trial of more than two years. Lydney Town lost its passenger service when the Severn Bridge was severed in 1960, but the Berkeley Road–Sharpness service survived four more years.

Coleford lost its passenger link from Monmouth when the section closed completely in 1916, but the service between Monmouth, Usk and Pontypool Road survived until summer 1955, being withdrawn a fortnight ahead of schedule because of an ASLEF strike. Monmouth–Usk closed completely, but Usk remained on the railway map until 1965 when the branch was cut back to its present stem to an Royal Ordnance factory at Glascoed, opening during World War II, with a four-platform station for workmen's trains, mostly from the South Wales valleys.

HEREFORD ROSS & GLOUCESTER

A secondary line on weekdays, the Hereford Ross & Gloucester was often used by North-to-West expresses diverted on Sundays when the Severn Tunnel closed for maintenance. The branch was the truncated outcome of the Monmouth & Hereford Railway incorporated 1845 and purchased by the GWR a year later, for a 22-mile line from Standish Junction (also the starting point for the South Wales Railway, projected at the same time) across the Severn at Framilode, with a 4½-mile branch from near Mitcheldean to the Forest of Dean Railway, and another branch to Grange Court. The Admiralty refused to allow the river to be bridged, claiming it would obstruct shipping. Little more was heard of the M & H after the Gloucester & Dean Forest Railway was authorised in 1846. Five years later the HR & G Company was formed with 'Head Offices' at Ross, to build a broad gauge line from Grange Court to Hereford (Barrs Court) and use the station jointly with the Shrewsbury & Hereford. The GWR provided financial support through the G & DFR. Construction took several years because of deep cuttings, four tunnels, and four crossings of the Wye. The southern five miles from Grange Court to a temporary station at Hopesbrook opened 11 July 1853, followed by the remaining 17½ miles on 2 June 1855. The GWR absorbed the local company in 1862 and in August 1869 used the branch as a guinea-pig for its gauge conversion programme, learning many lessons that proved useful for the rest of the work. Conversion to standard gauge enabled the GWR to introduce a goods service between Bristol, the Midlands and Birkenhead, using the Midland main line between Bristol and Standish Junction.

When Gloucester–Hereford passenger services were withdrawn under the Beeching Plan in November 1964, Hereford – Ross closed completely, the remainder following a year later together with the Monmouth branch south from Ross to an AEI siding at Lydbrook Junction.

WYE VALLEY AND LEDBURY BRANCHES

The ROSS & MONMOUTH, a twisting single line of tourist delight, was authorised in 1865, but the 12½-mile line from Ross-on-

Wye (to give that delightful town its full title—more than the railway
did), to Monmouth (May Hill) was not completed until 4 August
1873. It became a through route when the GWR opened a ¾-mile
extension to Monmouth (Troy) on 1 May 1874. It was worked by
the GWR from the start, although retaining nominal independence
until 1922. Despite its scenic attractions, the line was lightly used by
passengers, partly because the tourist season was short—and
few people lived in the area. A TUCC Inquiry shortly before
closure showed that the line was losing £10,000 a year, while local
buses, although carrying 300,000 passengers a year still could not
pay, their annual loss then being £1,500.

Passengers services were withdrawn on 5 January 1959 and the
seven miles north to Lydbrook Junction closed completely. On the
same day passenger services (losing £13,000 a year) were withdrawn
on the other arm of the Wye Valley route over the 13¾-mile branch
which joined the South Wales main line at Wye Valley Junction,
east of Brunel's bridge. It had been built by the MONMOUTH &
WYE VALLEY RAILWAY, opened 11 August 1876, ten years and
a day after authorisation. The line remained open to freight until
1964, when it was cut back to Tidenham Quarry (2½ miles from
Wye Valley Junction), which produces a particularly hard Dolo-
mitic limestone ideal for HST trackbeds. Further north part of the
trackbed has been converted into a pleasant walkway and Tintern
station attractively restored. See *Forgotten Railways: South Wales*.

The Gloucester–LEDBURY branch grew out of a scheme which
would have provided Ross-on-Wye with yet another railway. The
Ross & Ledbury of 1873 eventually built only 4¾ miles between
Ledbury and Dymock. It abandoned the remainder of its route to
Ross, and linked up with the Newent Railway, incorporated only a
few days after it, to construct a 12¾-mile line from Gloucester
(Over Junction) to Dymock. Both lines opened 27 July 1885.
Until the GWR opened the Birmingham & North Warwickshire
and associated lines, the branch provided the company with its
shortest goods route between Birmingham and Gloucester.
Gloucester–Ledbury passenger services, latterly worked by GWR
diesel railcars, were withdrawn in summer 1959 when Dymock–
Ledbury closed completely, the remainder of the branch surviving
another five years.

CHAPTER IX

From Paddington to Birmingham

Few routes have been more transformed in recent years than the two which the Great Western built between Paddington and Birmingham. The Aynho cut-off through the Chilterns has been downgraded and Paddington–Birmingham expresses, which now use New Street Station, once more take The Great Way Round to serve Reading and Oxford. It is a route seventeen miles further than Euston–Birmingham, and trains take up to an hour longer, but they meet a need. There has been gain as well as loss and Reading–Oxford–Birmingham has been improved to give Inter-City image and timings to north–south trains that have replaced one-time secondary route wanderers like the Pines Express via Bath and the Somerset & Dorset.

A new breed of Inter-City services has been evolved including a Paddington–Oxford–Birmingham–Manchester (Piccadilly) link, using the Leamington Spa–Coventry branch to serve Coventry and Birmingham International. Since May 1979 there has been a twice-daily service between the North-West, Gatwick and Brighton, running seven days a week, being aimed at air travellers. The curtain-raising special trip from Manchester on 3 April 1979 was for a demonstration flight on a Laker DC10 airliner. It was five days old, with decor in sharp contrast to the elderly, if comfortable Mark 1 coaches which made up the special train.

The Paddington–Birmingham routes were not among the most glamorous of the GWR, passing through quiet rather than spectacular countryside, but they were tremendously important, forming arteries especially busy with non-passenger traffic at nights: parcels and freights from Paddington and Birkenhead, iron-ore trains for Wrexham, joining the route at Banbury, freight to and from the Great Central, milk empties from Marylebone to Shrewsbury (Abbey Foregate), freight from the West Country to Crewe, freight from Park Royal to Bordesley Junction, newspaper trains from Paddington—a list almost endless.

Oxford became a tremendous operating bottleneck, especially in daytime when north–south expresses were changing locomotives. And inter-regional services do not always find favour with Oxford station announcers. A year later I got off the same service to hear a twenty-minute late arrival blamed on 'operating delays in the Midland Region.'

THE OXFORD RAILWAY

Maybe Oxford is still a little suspicious of railways. Certainly there are few places where they met with such strong opposition. The GWR included a 9½-mile Didcot–Oxford branch as one of three 'probable' ones in its original prospectus of 1833, with an extension to Worcester. A terminal planned at Magdalen Bridge, Oxford was opposed by Christ Church, which owned the land. It meant that for some years Oxford's long-distance travellers had to travel ten miles by coach to Steventon station. The 1½-hour journey was not popular, a factor which helped the Oxford Railway Company, solely GWR supported, when it was authorised in 1843. The University originally rejected railways because it felt that they would corrupt students' morals. Now the University had protective clauses written into the Act so that it could decide which students could travel.

The Company was absorbed by the GWR in May 1844 and the branch opened on 12 June to a small wooden terminal, more remote from the City centre than one at Magdalen Bridge would have been. A more spacious station of 1852 was rebuilt in the early 1970s. The original station, which had an overall roof, survived for goods until 1872.

By then there had been a further clash with the city caused by GWR plans in 1865 to establish its carriage and wagon (as opposed to locomotive) works on Cripley Meadow, behind the station, which the Corporation was wanting to lease to increase

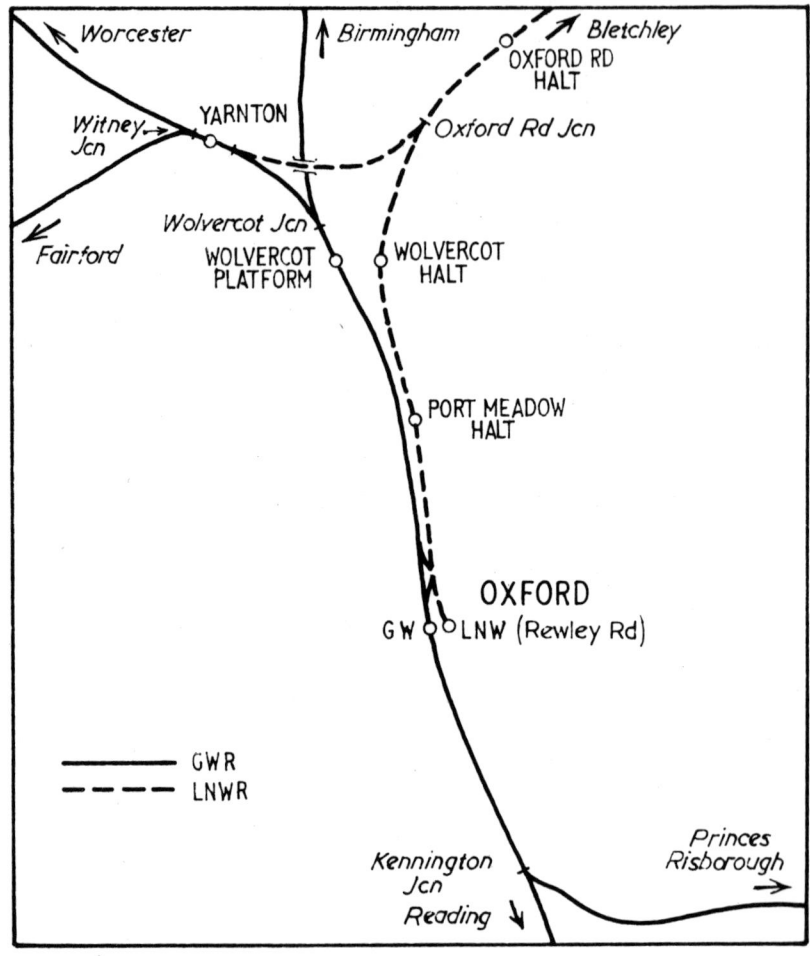

local trade. Oxford was preferred to Swindon because it had broad and narrow gauge lines while Swindon still had only the broad gauge. Other towns campaigned for the works: Reading, Abingdon, Banbury and Warwick, but at Oxford 'gown' rejected the proposal without compunction, and was supported by *The Times*, which felt that the works would disfigure an ancient and beautiful city and damage the interests of a national University.

Another critic, Goldwin Smith, wrote to the *Daily News* on 1 June:

The mass of the tradesmen in the City would lose more by the departure of a hundred students or residents connected with the University than they would gain by the arrival of five times the number of Great Western workmen.

Undeterred, the GWR went ahead preparing the lease with the Corporation, only to rescind it when Daniel Gooch became chairman in place of Richard Potter. Three years later the carriage and wagon works were approved at Swindon, costing initially about a third of those planned at Oxford even though, according to one critic, they were to have been built in the 'worst quarter' of Oxford.

The railways turned Oxford into a tourist centre and the GWR exploited the trade to the full. In the 1920s it offered a set of 107 lantern slides free to meeting organisers throughout the country. Called 'Oxford. A walk round England's greatest University City,' it was complemented by a shorter one on the Thames Valley.

The development of railways north of Oxford was largely influenced by the demands of Midland industrialists for lines to break the London & Birmingham's monopoly, and by growing agitation and need for a link-up of already-established main lines of the Midlands and Southern England.

OXFORD & RUGBY

The GWR responded by promoting the Oxford & Rugby Railway, which ran immediately into the expected opposition of the L & B, which put forward the London Worcester & South Staffordshire Railway. It was designed to protect the Birmingham company's existing territory and keep the broad gauge out of a wide area. It was projected from Tring through Aylesbury, Bicester, Banbury, Fenny Compton, Evesham, Worcester, Stourport and Dudley to Wolverhampton. Oxford was to have a branch from Bicester, Rugby one from Fenny Compton. The branches would have directly challenged the O & R, especially as an Oxford–Didcot extension was planned to join the London & South Western's projected Basingstoke & Didcot Junction Railway, later abandoned. When the industrialists suggested a route from Oxford through Worcester to Wolverhampton, the GWR put forward a more direct route from the O & R north of Banbury.

Many schemes were considered by the Board of Trade and the eminent men who became tagged as 'The Five Kings,' and in January 1845, at the start of the Railway Mania, they reported against any broad gauge extension, favouring the London & Birmingham scheme since it would prevent the GWR from having a monopoly stretching from London to South Wales and Birmingham.

But Paddington pressed ahead and two lines—the Oxford Worcester & Wolverhampton and O & R—both promoted through subsidiary companies. They were authorised together on 4 August 1845, although broad gauge aspirations were dampened by a stipulation that narrow gauge rails must be added to the Rugby line, and to parts of the Wolverhampton route, if the Government called for them.

BIRMINGHAM & OXFORD JUNCTION

A year later, before construction of the O & R had begun, the Birmingham & Oxford Junction was incorporated on 3 August 1846. It was to run from the O & R at Knightcote, two miles north of Fenny Compton, to Birmingham and a junction near Curzon Street with the LNWR, *officially* constituted only a few days earlier. Events in the gauge battle moved fast and the B & O, together with the Birmingham Wolverhampton & Dudley, authorised at the same time, soon agreed to amalgamate, and on lease or sale to the GWR since it was considered likely, if not certain, that both lines would be broad gauge.

The gauge question led to a rash intervention which turned out to be an embarrassment for Mark Huish, who as secretary of the Grand Junction Railway was arch-enemy of the London & Birmingham. When that company opposed the GWR plans, Huish rushed to the aid of Paddington and advocated an alternative broad gauge route from London to the Midlands and possibly north to the Mersey. His concept of a line between Birmingham and Oxford was one he forgot when the L & B and GJ were involved in the formation of the LNWR, of which be became general manager.

Because of the Gauge Commissioners' Report, none was stipulated in the B & O Act, and moves by it and the BW & D towards union with Paddington, provoked Euston to embark on a shady and historically famous take-over bid for the B & O. Huish obtained four-fifths of the shares, but opponents demanded a Parliamentary Inquiry into the dealings, and it led to the lease of both companies to the GWR being confirmed.

The case for narrow gauge was strongly pressed by Euston, largely to irritate Paddington and force it to spend precious capital. For the LNWR had little if any intention of exchanging traffic with the B & O at Curzon Street, Birmingham, Leamington or Banbury, despite blandishments by Huish.

The B & O gauge question was reviewed after the Government insisted on the addition of the narrow gauge, forcing the reluctant GWR to consider Oxford–Basingstoke as a mixed gauge route. Later it agreed to make a narrow gauge line after Parliament rejected a scheme for the Manchester & Southampton Railway, essentially a narrow gauge line between Cheltenham and Southampton. The GWR was by no means getting its own way in developing a line towards the Midlands.

Two schemes which could have very much altered the final shape of the railway map are worthy of short note: the London & Oxford of 1846 was designed to breach broad gauge territory and make sure of the establishment of the narrow gauge west to Hereford. It was to have left the Birmingham line near Harrow and run via High Wycombe, Oxford, Burford, reaching Hereford through Ledbury. The company dissolved when the House of Commons rejected the Oxford portion. The other narrow gauge line that failed was to serve Oxford, Witney, Cheltenham & Gloucester, but its broad gauge rival actually came into being: the Cheltenham & Oxford, authorised 9 July 1847, to run 37½ miles from the GWR at Cheltenham to the O & R near Rugby, with a half-mile branch to Witney. This project was forgotten once the OW & W went ahead but later, very much in its footsteps came the Banbury–Cheltenham cross-country route. The Cheltenham & Oxford had shown a need, albeit prematurely.

CHANGE OF DIRECTION

Construction of the Birmingham & Oxford was slow for several reasons including awkward landowners, and early in 1849 Paddington decided to head for the Midlands rather than Rugby, where gauge interchange, apart from being unpopular, would have been a handicap to through traffic. The 15½ miles from Knightcote were abandoned. By then the railway system had been altered by the Buckinghamshire Railway, an LNWR line of 1847, formed by the amalgamation of two companies through which Euston was opposing the B & O. They were the Buckinghamshire & Brackley and the Oxford & Bletchley. A Banbury–Aylesbury branch was never built.

The first section of the B & O, a single line of purely broad, rather than mixed gauge, was opened between Oxford and Banbury on 2 September 1850—a 24-mile stretch that ran to Millstream Junction, Oxford, ¾-mile south of the Oxford branch terminus. It meant that trains had to reverse to reach it until doubling in 1856, when mixed gauge working started between Oxford and Basingstoke. There were only three intermediate stations between Oxford and Banbury, Woodstock Road, (Kirtlington, when a new Woodstock Road opened two miles nearer Oxford) Heyford and Aynho.

The rest of the line to Birmingham, opened as a double mixed gauge route on 1 October 1852, incorporated a major deviation at Leamington Spa. It had been authorised in 1848 to take the line through, rather than to the west of the town. Euston was approached about construction of a short joint stretch, but after talks broke down each company was forced to build its own embankments and viaducts. After rationalisation in the 1960s, sections of the LNWR were demolished, but ugly abutments remain.

The B & O took its line into Birmingham on a spacious scale and several modifications were made after relations with Euston improved (Vol 7).

North of Banbury there were nine intermediate stations—a total almost doubled in this century by seven more in the Birmingham area. Oxford, Banbury and Leamington Spa were distinguished by Brunel over-all roof stations, though mixed gauge trains could not use Leamington because the platforms only had broad gauge rails, and a dowdy mixed gauge station had to be built on a loop to the north-east. Following opening of the Oxford Worcester & Wolverhampton in 1853, the mixed gauge was brought into use between Oxford and Wolvercot, three miles north.

Gauge problems plagued operating for years and the GWR did

Plate 29

Above: One of the major operating obstacles on the Midland route from Bristol to Birmingham was the 1 in 37 Lickey Incline between Bromsgrove and Blackwell. Midland class 2 4-4-0 No 537 coasts through Bromsgrove at the foot of the incline with an express from Derby to Bristol, including an LNWR through coach, second from the front. (*L&GRP, courtesy David & Charles*)

Plate 30

Below: Nearly all trains climbing the incline, except for the lightest, required banking assistance, often with two locomotives or the famous 0-10-0 'Big Bertha'. Here a coal train breasts the summit at Blackwell with two 0-6-0 tanks pushing at the rear. (*L&GRP, courtesy David & Charles*)

Plate 31

Above: The area covered by this book abounded with principal cross-country main lines or cross-country services. Around the first world war period, Badminton class 4-4-0 No 4112 *Oxford* heads the Newcastle–Bournemouth train, composed mainly of LSWR stock with a GCR coach next to the engine, past Kings Sutton, junction for the branch to Cheltenham, as it runs south for Oxford. (*L&GRP, courtesy David & Charles*)

Plate 32

Below: A down local train from Oxford to Banbury arrives at Kings Sutton headed by 2-4-2T No 3630. (*L&GRP, courtesy David & Charles*)

not consider through narrow gauge services until after it absorbed the Shrewsbury companies in 1854. It was another two years before narrow gauge workings were introduced between towns north of Wolverhampton and the South via Basingstoke. The problem disappeared when the broad gauge was abolished north of Oxford on 1 April 1869. Conversion mainly involved eighty mixed gauge miles between Oxford and Wolverhampton.

THE LINE TODAY

In more recent times, the three main stations have been rebuilt. Those who knew the striking, white Leamington Spa station at the time of its creation in 1938 with its walnut veneered waiting rooms, windows edged with stainless steel, and the long-wide and widely-separated up and down platforms outside, will find it virtually untouched today. It is a splendid place where the mind can automatically put back the clock for many passengers—as I found when I visited it after an absence of more than thirty years. Another station rebuild, this time in the years just after rather than before World War II, has been Banbury, which does not handle as many of north–south services as Leamington. A station built in a straight-forward style, it lacks the character of Leamington.

Continuous station change is taking place at Oxford, where a much-needed, long overdue new station was officially opened on 1 August 1972. It has the same 'pre–fab' ambience as Bristol Parkway, is bright and cheerful, with a silver-walled subway, and provides comfort and electronic train information for passengers. An ambitious plan provides for station enlargement with a new booking hall and other facilities. Submitted by BR and Associated Dairies to Oxford City Council in 1979, it was for comprehensive redevelopment of land outside the station, which had been ear-marked for a road scheme, now abandoned. The planners want a bus-rail interchange, a large hotel and supermarket and residential and industrial development on the site, including that of the LNWR station, Rewley Road, which was listed for demolition.

Modernisation has brought many changes to the Birmingham & Oxford, apart from the services mentioned at the beginning of the chapter. Shortly before the closure of the Great Central London

Extension and Banbury branch in 1966, a new junction was opened at Leamington Spa (15 May) between the Birmingham and Coventry lines, to handle diverted traffic. Paddington–Birmingham–Birkenhead expresses were withdrawn on completion of Euston–North-West electrification, and most Paddington–Birmingham Inter-City re-routed via Oxford from May 1973.

The Western Region regards Oxford as a railhead for commuters, and when you see a crowded evening rush hour train from Paddington arrive at Oxford you understand why. A BR leaflet *Railways for Cities* showed Paddington–Reading–Oxford as among lines for electrification sometime in the future, and a similar proposal was contained in Sir David Barran's London Rail Study of 1975, listing priorities up to the year 2000.

But BR has to compete with the motorway lobby for available money, and the battle of the 1980s is likely to be over extending the M40 between Oxford and Birmingham, estimated to cost £100 million when it was first announced on 23 January 1978. The Western Region contends that if the railway system were encouraged to carry an extra six million passengers and six million tons of freight in a comparatively small and close-knit area, the motorway would be unnecessary. It says that there is plenty of spare capacity on London–Birmingham routes, both via Oxford and Rugby, while the High Wycombe–Oxford–Banbury 'corridor' could handle an extra 2,000 passengers and 7,000 tons of freight daily.

GREAT CENTRAL BANBURY BRANCH

Two new lines had a profound influence on the B & O in the first decade of the present century, the Aynho cut-off, which altered the express pattern, and the Great Central Banbury branch of 1900, used by a number of long-distance express and freight services. Branch is a misnomer for this short line, as it was effectively a main line more important to the Great Central and its successors than the London Extension, which despite Marylebone dining car expresses was in its final years little more than a duplicate route to London from South Yorkshire, Nottingham and Leicester.

The 8¼-mile double track line between Culworth Junction, near

Woodford and Banbury was built under GWR pressure and with that company's money, the Act of 1897 enabling Paddington either to lend the capital to the GCR (then engaged in a financial struggle to reach Marylebone) or to build the branch itself if necessary. In the event the GCR did so, and *The Railway Year Book* noted it as an 'important link between the North East and the West of England.' It opened to goods on 1 June 1900 and to passengers from 13 August. Long-distance express services soon developed, but the branch is remembered as one of freight exchange. In 1940 it was used by trains totalling almost 700,000 wagons, while a decade later Banbury was the busiest freight interchange on the Western Region. The line closed, together with much of the Great Central, on 5 September 1966, followed by the reduction to terminal sidings of the GC Exchange Yard (with the displacement of eleven men) on 4 October 1971.

The economies were another stage in decline which robbed Banbury of important junction status after more than a century. It had included the loss of Chipping Norton passenger services from 4 June 1951 and of a spur to the LNWR Banbury–Buckingham line, closed with that branch 6 June 1966. Another Banbury casualty had been the GWR locomotive shed just south of the station, which had an allocation of some eighty locomotives (about twenty-five more than Oxford)—a place where even Kings sometimes lingered.

TWO GWR BRANCHES: WOODSTOCK AND ABINGDON

Road competition led to the total closure of the Fairford branch (page 101) and that from Kidlington to the ancient borough of WOODSTOCK, eight rail miles from Oxford, three-and-a-half from the Birmingham main line at Kidlington, although local services—auto-trains of one or two coaches—used their own line beside the B & O for a mile. The intermediate halt, Shipton-on-Cherwell, was one-and-a-half miles short of the terminus, which for many years was called Blenheim & Woodstock.

Blenheim is the seat of the Dukes of Marlborough and in late Victorian years the Duke financed the branch, constructed by the Woodstock Railway, which opened 19 May 1890, the work having taken two years. The GWR bought the company for £15,000 a few weeks afterwards. Some trains ran through to Oxford, and a few were mixed to cope with light goods traffic. Passenger services were withdrawn and the line closed completely 1 March 1954, a man of 86 who helped to build the branch being a passenger on the last train. Woodstock station subsequently became a filling station, but the branch is commemorated by a model in Pendon Museum near Didcot.

A string of wayside stations survive on the Western Region 'Commuteroute' tip between Didcot and Oxford. They include Radley, once the junction for ABINGDON branch passengers. After the town rejected a branch from the projected Didcot–Oxford line of 1837, nothing was done for seventeen years until a local company was formed to build a line of just under two miles to near Culham. It opened on 2 June 1856 and in 1873, a year after conversion from broad to standard gauge, it was extended for ¾-mile to Radley, where a new station was opened. The branch passenger services were withdrawn in September 1963. They had never been popular because of the change necessary at Radley. The branch survives for freight traffic of Associated British Maltsters.

THE WYCOMBE RAILWAY

For years there was an alternative passenger service between Oxford and Paddington to the direct one via Reading. In his evocative book *Gone With Regret*, George Behrend remembered how passengers arriving at Oxford station for a London express could be dispatched into a suburban train built for the City services and simply labelled Paddington, which took a circular tour of the Home Counties, often via Bourne End rather than direct between High Wycombe and Paddington. The trains that ran via Bourne End covered virtually all the Wycombe Railway, one of those rare companies that stretched beyond the area suggested by its title, eventually completing thirty-seven miles between Maidenhead and Oxford and another seven to Aylesbury.

The original ten miles from Maidenhead to Wycombe, authorised in 1846, were delayed by financial difficulties and not opened until 1 August 1854. The junction at Maidenhead was 1½ miles west of the original station, and when the branch opened it was leased to the

GWR. It was broad gauge, single and cheaply built. In 1857, a fourteen-mile extension from High Wycombe through Princes Risborough to Thame was authorised and it opened 1 August 1862. A year earlier the last thirteen miles to Oxford (Kennington Junction) from Thame was authorised and the route was completed 24 October 1864. Both stretches were single and broad gauge, but only for a comparatively short time, the whole of the Wycombe Railway being converted in 1870.

Completion to Oxford had been vigorously opposed by the West Midland Railway, still chaired by William Fenton, former chairman of the OW & W. He contended that Thame–Oxford was neutral territory which neither company must invade. It was also seen as a direct threat to its own plans to reach London with the London Buckinghamshire & West Midland Junction Railway from Yarnton via Thame, Princes Risborough and Amersham. Aylesbury was to have a branch. The scheme was forgotten from the moment the GWR got control of the West Midland.

The Thame route was one on which the GWR followed the LNWR example of three years earlier by introducing a steam railcar service. It began in 1908 to try to stimulate Oxford suburb traffic and halts opened at places were there was the possibility of traffic. One, beside the Thames, was to serve Iffley, but a towpath, lock, toll-bridge and more than half-a-mile separated platform and village. Sometimes the footpath was flooded, but as Kingsley Belsten noted in the *Oxford Times*, there was an alternative for stranded passengers—a footpath across the main line to Kennington, and a pub! Another halt, Garsington Road, closed when the rail-motors were withdrawn during World War I, was revived in 1928 as the Morris and Pressed Steel works expanded. When Morris Cowley station closed in 1963 it was, reported Belsten, taking more in car parking fees than in fares.

Oxford–Princes Risborough regular passenger services were withdrawn in January 1963, just ahead of Beeching, and Morris Cowley–Thame closed completely in May 1967. The A40 over-bridge at Wheatley was later used to bridge a river on a diverted track stretch at Cwmbach, near Aberdare. Bourne End–High Wycombe closed to goods in 1967, the passenger link surviving between Maidenhead and Aylesbury until May 1969 and to High Wycombe a year longer. Maidenhead–Bourne End continues to

provide access to the Marlow branch while Princes Risborough–
Thame survives for oil traffic, while at Oxford freight still runs
between Kennington Junction and the BL Works. Faced with
growing traffic problems, Oxford City Council and the County
Council have considered reviving a passenger service to the Works.

PRINCES RISBOROUGH : COUNTRY RAILWAY CENTRE

In just under half a century from the arrival of the Wycombe
Railway, the small and pleasant town of Princes Risborough
developed into a busy railway centre, for besides the routes to
Oxford, the Great Central and Birmingham, it was the junction of
two branches. First to open was the 7¼-mile AYLESBURY branch,
authorised with the Thame–Oxford extension, which was completed
broad gauge to a Joint station with the Aylesbury & Buckingham
Railway on 1 October 1863. It was one of the last and shortest-lived
broad gauge lines, being converted in 1868, a year after the GWR
absorbed the local company. It was Aylesbury's second railway,
the London & Birmingham having arrived from Cheddington in
1839. The town got its most direct line to London in 1892 with the
Metropolitan route of thirty-eight miles, five less than the route to
Marylebone via Princes Risborough. In 1907, the Princes Risborough–Aylesbury line had a second change of ownership, being
taken over by the GW & GC Joint Committee, giving that system a
final length of forty-one miles. Despite the mileage handicap, the
branch has several Marylebone through trains Mondays–Friday.

The 8¾-mile Princes Risborough–WATLINGTON branch once
offered something special to tired businessmen escaping to some of
the loveliest and quietest countryside within easy distance of
London, a slip coach provided until 1956, off the 7.10pm Paddington
–Birmingham express—every night in summer, Fridays and
Saturdays in winter. It was slipped for attachment to a Watlington
train which, after leaving the station, ran for nearly ¾-mile alongside
the Oxford line before turning to take its own course under Chiltern
escarpment. The Watlington & Princes Risborough Railway was
promoted locally by a private company. It took the initiative once
it became clear that Wallingford & Watlington Railway (page 29)
would never cross the Thames, and the line of six intermediate
stations and halts, opened on 15 August 1872. It ran to a terminus

nearly a mile short of the village, but the local trains, which had once been formed of railmotors, survived until beaten by bus competition in 1957. Four years later the branch was cut back to its present length of almost four miles, kept open to serve a large cement works at Chinnor.

THE OXFORD CUT—OFF

The GWR route between Paddington and Birmingham by way of Bicester is rapidly popularizing many delightful places both as travel and residential centres.
It must be noted that the line through Denham and Gerrard's Cross to Wycombe, part of the GWR route to the Midlands and the North, serves a district of unrivalled beauty . . .

Quotes from *Holiday Haunts* 1921 reflecting the pleasures of the 'gigantic Oxford avoiding curve' as W. J. Scott described the Ashendon & Aynho Railway in the *Railway Magazine* for June 1905.

Yet it was only part of an historically complicated route developed so that the Great Western could finally rid itself of the tag of The Great Way Round. This was achieved by the Badminton Line of 1903, the Berks & Hants of 1906 and Ashendon–Aynho of 1910. The latter line reduced the Paddington–Birmingham distance by nearly nineteen miles and gave the GWR a route some two miles shorter than that of the LNWR via Rugby.

It was created in stages, partly by itself and partly with the willing help of the Great Central Railway. The springboard was a 23-mile line from the Bristol main line at Acton (Old Oak Common West) to the Wycombe Railway at High Wycombe, with a Loop from the main line at Ealing to Greenford (Vol 3). It was authorised in 1897 when the Great Central was within two years of completing its London Extension and was unhappy at having to use forty miles of the congested and steeply-graded Metropolitan Railway between Quainton Road and Neasden to reach its new terminal at Marylebone. The Great Central, already working closely with the GWR over the construction of the Banbury branch, saw a far better prospect in the GWR line, although it would be $4\frac{1}{2}$ miles longer to Marylebone via a connecting line from Northolt Junction. And there was the promise of a share in suburban traffic.

GREAT WESTERN & GREAT CENTRAL JOINT COMMITTEE

Agreement with the GWR led to the formation of the Great
Western & Great Central Joint Committee on 1 August 1899, to link
the systems with lines old and new. Besides taking-over construction
between Northolt and High Wycombe, the Committee purchased
from the GWR 8½ miles of the Wycombe Railway between High
Wycombe and Princes Risborough to double and improve it, and
was authorised to build fifteen miles of new line between Princes
Risborough and the GCR London Extension at Grendon Underwood Junction, nearly three miles north of Quainton Road. The
last two projects were to provide the GWR with the second part
of a direct route to the Midlands.

The Wycombe stretch was improved out of all recognition and the
whole line laid out to cope with express and local traffic. New
stations were built at High Wycombe and Princes Risborough,
places which were brought closer to the Capital, the direct line being
six miles shorter to High Wycombe than via Maidenhead.

During construction of the Joint Line, the companies agreed that
the northern end should be at Ashendon rather than Grendon
Underwood Junction, and the GCR took over construction of the
intervening six miles. Northolt Junction–Ashendon Junction,
almost thirty-four miles, opened to goods 20 November 1905 and to
passengers 2 April 1906, with local services to Paddington and
Marylebone and GCR expresses via Grendon Underwood. GWR
passenger trains did not run between Princes Risborough and
Ashendon until the Aynho route opened in 1910.

Suburban traffic over the Joint Line never developed to the
extent that the companies hoped—the Great Central had perhaps
more to gain than the GWR since the express route was the important thing, and the Great Central was optimistic in its statement
in 1914 that 'It opens up a large residential area and a considerable
suburban traffic has been created by providing fast and convenient
trains.' It lay on the fringe of Metro-land, which grew faster because
it was a little closer to London, and the pattern has remained to the
present day, BR and London Transport services being more
intensive than those on the High Wycombe line.

Until modernisation of the 1950s and 1960s the Joint Line retained
its two-company character, which made it one of the most exciting

of the London suburban routes, for from stopping trains made of ex-GC 4–6–2 tanks and a string of teak compartment coaches, one could glimpse GWR King class locomotives thrashing along 'middle road' with the Birmingham two-hour expresses, or the long-distance Great Central expresses, which for years had ex-GC express locomotives at their head. Stopping trains took just over an hour between Marylebone and High Wycombe, under thirty miles. And on ones like the 5.20am from Marylebone (caught after an overnight journey from the North West to Euston), the journey seemed interminable. But at least they were memorable.

ASHENDON & AYNHO RAILWAY

The Joint Line provided the GWR with a Paddington–Oxford line eight miles shorter than via Didcot, but it was never considered practical for expresses because of its single, twisting nature west of Princes Risborough. The GWR wanted a Birmingham direct line, which came with authorisation in 1905 of an 18-mile line from Ashendon through the Chilterns to Aynho, five miles south of Banbury.

The Aynho cut-off involved heavy engineering works with the up and down lines carried for 2½ miles through the Chilterns on split levels in deep chalky cuttings. Bicester, 'known for its ale and pillow lace' noted the 1875 *LNWR Guide*, was the most important of five intermediate stations. Bicester North became the small town's second station, but travellers mindful of changing to the Oxford–Bletchley branch were warned that London Road station was more than a mile away. Brill & Ludgershall station lay close to one on the Brill branch, which lost passengers as a result. It also suffered when the Great Central opened a station at the intersection of the branch and its Ashendon–Grendon Underwood route.

The cut-off opened to goods on 4 April 1910 and to passengers on 1 July. Express drivers took full advantage of the flying junctions at Ashendon and Aynho that had been well proportioned for fast running, although Aynho was not built as once envisaged. According to W. J. Scott in the *Railway Magazine* for June 1905, there was to be a loop curve at Aynho North Junction facing towards Adderbury to connect with the Banbury & Cheltenham line, 'which—if a "straight road" avoiding Chipping Norton

Junction be afterwards added—would give a good alternative third route from London to Cheltenham.' It was also suggested, though not by Scott, that expresses would use a mile-long new line between Aynho and the Cheltenham line at King's Sutton to keep them clear of the Banbury–Oxford tracks. Chipping Norton got its avoiding line, but the years never produced a need for a third route to Cheltenham.

Paddington–North West expresses were withdrawn from the direct line in 1967 and the following year, twenty-six miles from just north of Princes Risborough to Aynho were singled. Three signal boxes closed when the stretch came under the control of the North Box at Princes Risborough, where stopping trains were concentrated on the 'town side' of the station to get rid of a footbridge.

The Joint Line is now designated on BR Passenger Network Maps a thin red line—not Inter-City. Commuter services are based solely on Marylebone. Yet when diesel multiple-units were introduced in 1962, there were services to Paddington as well, and for a time a plan was considered to close Northolt–Neasden and switch all trains to Paddington. It was supported by some commuters disgruntled when the Western Region handed-over the services to the London Midland. Some Marylebone trains run through to Banbury (68¾ miles), though most turn round at High Wycombe or Princes Risborough. The pattern could change again if BR goes ahead with a plan of the late 1970s for another cross-London route linking the Midland and Southern Regions by re-opening Snow Hill tunnel in Central London. The plan envisaged an interchange at West Hampstead for Aylesbury, High Wycombe and other commuter services. Meanwhile, a single Paddington–Birmingham express is retained in each direction.

OXFORD: LONDON & NORTH WESTERN RAILWAY

The importance of GWR–LNWR connections at Oxford increased through the years. Euston maintained a small but distinctive presence at its own terminal, Rewley Road, a car park's width from Oxford General station. Ironwork from the Great Exhibition in Hyde Park was used in construction of Rewley Road, and if the building, now a tyre store, is demolished, the ironwork is scheduled for museum preservation. The terminus, closed 1 October 1951 when the Oxford–Cambridge 'Brain Trains' were diverted to Oxford General, was built by the Oxford & Bletchley Junction Railway of 1846, an arm of the Buckinghamshire Railway. The 29½-mile line reached Oxford 20 May 1851, approaching from the North alongside the Birmingham line, opened the previous September as far as Banbury. The line provided a valuable link through to Worcester once the Yarnton Loop opened in 1854 (page 88). The GWR let the Bletchley branch trail off its maps without stating where it was going, obviously to make it seem inconsequential, but to other companies it was important. The Great Eastern noted Yarnton in its *Guides* as a route to Worcester and Eastern England.

The line became important as a through route, especially for freight avoiding London, a fact emphasized in 1940 when a south-facing connection was put in with the Birmingham & Oxford, just north of the station, and used from 8 November. The branch came into renewed prominence in the mid-1950s when planners were developing through freight routes, again to keep the congested London network as free as possible of such traffic. Associated with the branch was the Bletchley fly-over, which soon fell into disuse as rail freight dropped in volume and services were rationalised, BR finding itself with a lot of spare capacity, especially as the liner train concept led to the closure and abandonment of marshalling yards; there was to have been a new one on the Oxford–Cambridge route.

As part of the Oxford master automatic signalling scheme associated with the new station, the junction with the Bletchley line was moved north from Oxford, almost to Yarnton, where the loop had closed in November 1965.

Oxford–Cambridge passenger trains survived until 1968, some nine years after withdrawal was first mooted and five after Beeching, who recommended the closure of several small intermediate stations, though not the withdrawal of the through trains over the 77-mile cross-country route. In the final years ten daily trains took at best just over two hours. They were used by about 3,000 passengers a week at Oxford, 2,000 at Bicester. Recently there has been agitation for the restoration of passenger trains over the first twelve miles between Oxford and Bicester. Consultants to Milton Keynes New Town expressed interest, although they did not advocate an extension through to Bletchley.

Oxford–Bicester has never supported a frequent passenger service, even when Oxford's northern suburbs were growing in the days before road competition. The LNWR introduced steam railmotors in October 1905 and opened several roadside halts. The service was not well used and was a casualty first of World War I (1917—19) and of the General Strike in 1926 when it was finally withdrawn, local trains continuing to serve the long-established stations.

CHAPTER X

Through Routes

THE READING CONNECTIONS

The stem of the Reading connections discussed in this chapter is the Berks & Hants, which laid the foundations of the direct line from Paddington to the West. It began at Reading, debatably the most important main line junction that the GWR ever developed. It is still among the most important on British Rail, and while many railway centres have lost traffic through closures Reading has benefitted from the demise of cross-country routes and, as noted, the concentration of North–South through traffic via Didcot. It is a railway centre that has reasonable prospect of growth, for when the Channel Tunnel was being projected in the 1960s, Reading was being considered as one of the main terminals, fed from the electrified Tonbridge route.

The concept is not new: the South Eastern & Chatham rated Reading as a useful traffic exchange. In its *Official Guide* for 1907, it headed its 'East to West and North Services,' with: Bâle, Berlin, Cologne, Brussels, and Paris to Dover for Reading, and listed several other routes by which Continental travellers could avoid London. It included a 'Special express service from Dover–3 hrs 21 mins,' and stated: 'Sufficient time is usually allowed at Reading for passengers to partake of substantial refreshments before joining the trains which proceed northward through Oxford, Leamington, Birmingham and Wolverhampton to Crewe and Manchester; to Shrewsbury and Aberystwyth, Llangollen and Barmouth. And also via Chester to Birkenhead, Liverpool and Manchester.'

While its through service plans were grossly exaggerated on paper, it is worth quoting this *Guide* to show the aspirations of one of the big companies in the golden age of competition. The *Guide*

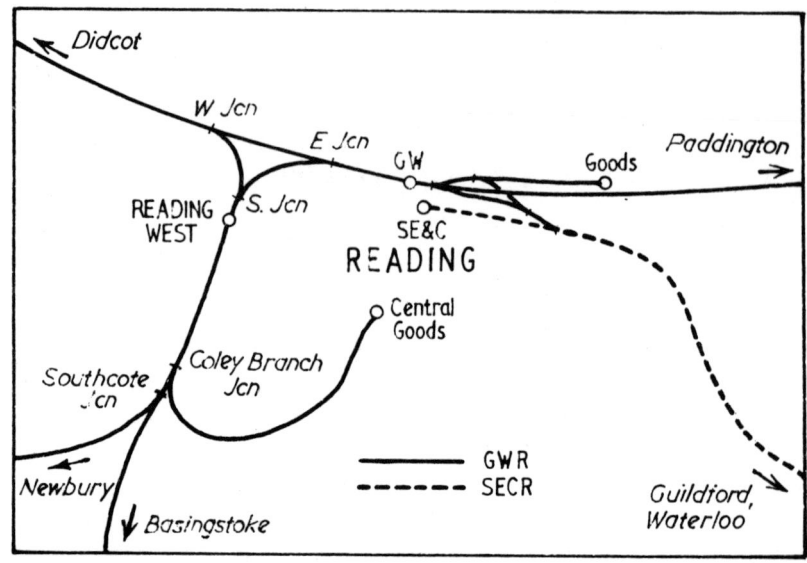

was published a decade after the start of a Birkenhead–Folkstone Continental service, among a group of which the longest survivor was to be Birkenhead–Dover, which for decades took Southern-green coaches into Birkenhead Woodside station.

Nowadays there are no booked passenger services to the Kent coast via Reading, but the route remains busy with freight. When London avoiding routes were considered in 1957, there were proposals to upgrade the Tonbridge Line and build a high-speed burrowing junction, and improve traffic exchange.

GWR DEFENCE OF READING

The link at Reading was forged in 1858 after the GWR had spent several years satisfying itself that the South Eastern was not wanting to use the connection to drive a standard gauge line through its broad gauge territory north of Reading. Nevertheless, the GWR established a track layout to discourage any liberties, insisting that the spur pass under the London main line to a junction lying east of its own Reading East Station Junction. This forced the South Eastern Railway to stop its trains in the GWR station. They had to set back into a wooden platform squeezed between the main station and the locomotive shed.

The present connection dates from wartime: 26 May 1941. It replaced one opened 17 December 1899, a year when Reading got a new passenger station in 'striking contrast to that opened in the early morning of April 4th 1840,' according to a GWR Guide. 'The station has three platforms, with bays which can increase the number of possible starting places to ten. The subway, with staircases and hydraulic lifts to each platform, is just 100 yds in length. Modern streets and *boulevards* of continental aspect now cover the open space' [outside].

The station approach was drastically altered after the Southern Railway's old station, Reading South, was closed on 5 September 1965, demolished and the site became a much-used car park. Electric trains to Waterloo (43½ miles via Staines against thirty-six to Paddington) use a new spur to reach a main line platform bay. It was used also by Redhill–Guildford–Tonbridge diesel multiple-units, which had to cross the main line to other platforms at peak periods. Such movements had to be eliminated on the birth of the HST age, and a second platform was opened 5 May 1975. It had not been constructed earlier because Dr Beeching had recommended withdrawal of the Tonbridge service.

Passenger and goods traffic have been well segregated at Reading, where the 1¾-mile Reading Central–or Coley–goods branch runs to a spacious depot beside the river Kennet and close to the town centre. Branching from the Berks & Hants immediately north of Southcote Junction, it loops round into the depot. Authorised in 1905 and opened 4 May 1908, it was built to main line standards and could accept every steam locomotive except a King.

BERKS & HANTS

The Berks & Hants has a special flavour, as I found when I travelled on the Cornish Riviera Express to produce a radio programme to mark its 70th birthday. As we swung off the Bristol main line at Reading, I became aware of transition from town to a countryside looking green and delicate on an early spring day. The next city was miles ahead: Exeter, whose heart lies within sight of fields rather than factories. The B & H is a glorious curtain-raiser to the delights of the West Country. It is still commuter land, though only just, enjoying a prosperity born of London's expanding residential

demands. Ten intermediate stations (plus that for Newbury Racecourse) survive in the 59½ miles between Reading and Westbury
although Pewsey is the only one in almost half that distance, the
twenty-nine miles between Bedwyn and Westbury. The only station
that qualifies for heavier black type in BR Timetables is Newbury—
a 'quaint old Berkshire town' to quote a GWR guide—now a
growing railhead having besides a good local service, an Inter-City
one, taking under the hour to Paddington.

The B & H was born of the Railway Mania and several schemes,
primarily for linking established lines in the South with those to the
Midlands. The nominally independent company was authorised
in 1845 with strong support from the GWR, anxious to thwart a
plan by the L & SWR to reach Didcot from Basingstoke.

The B & H was to build two broad gauge lines radiating from
Reading: south to Basingstoke (13½ miles) and west via Newbury
to Hungerford. After the Board of Trade supported the company
because its plans would close the north–south gap, the South
Western and GWR agreed not to penetrate each other's territory.
The following year the Gauge Commission strongly advocated a
narrow gauge link between Oxford, Reading and Basingstoke, but
it was another ten years before it was in operation, partly because
of the GWR's reluctance to mix gauges.

Meanwhile, the first 25½ miles of the B & H, between Reading
and Hungerford opened 21 December 1847, followed by the
Basingstoke branch from Southcote Junction, Reading, on 1
November 1848. The line, perhaps more important now than ever
before, has been under Southern Region control since 2 April
1950 (Vol 2). The Kennet Valley line was extended 24½ miles from
Hungerford to Devizes on 11 November 1862 by the Berks & Hants
Extension Railway, incorporated three years earlier. Its title made
railway rather than geographical sense for it never reached Hamp-
shire and ran only two miles in Berkshire. Wiltshire was the county
it really served, helping to develop the richly agricultural Vale of
Pewsey.

DIDCOT NEWBURY & SOUTHAMPTON

Between Reading and the West, large areas which the GWR had
tried unsuccessfully to keep as its own were penetrated by three

cross-country lines: the Somerset & Dorset, the Midland & South Western Junction, and the Didcot Newbury & Southampton, the last being the only one to be GWR-worked. After authorisation in 1873, the Didcot Newbury & Southampton Junction lay moribund before landowners saw it would meet local rather than wider needs and got a fresh Act in the summer of 1880. It included powers for GWR connections at Newbury (not in the original scheme of the 'Junction' company), and a junction with the LSWR about a mile nearer Micheldever Station. There was strong dissatisfaction in Southampton with the LSWR, and the DNS was promised support for an independent line from the North (Vol 2).

Political battles led to the DNS being curtailed to almost forty-five miles: 17½ miles across the Berkshire downs from the GWR at Didcot to Newbury, opened 13 April 1882 with four local stations, the company stating 'about 50 villages are thus accommodated,' and 27½ miles south from the Berks & Hants at Enborne Junction, Newbury, to Winchester (Cheesehill). That opened 1 May 1885. By then another section had been authorised covering 10½ miles from Burghclere, north-east to the Berks & Hants Extension at Aldermaston. It was abandoned in 1891 in what a respected historian, T. B. Sands, termed in the *Railway Magazine* February 1955, 'a final gesture of defeat.' It happened as a new LSWR connection (Shawford Junction) was about to be opened at Winchester.

The Aldermaston line was intended to provide a competitive Paddington–Southampton route and undeterred by its loss, the DNS developed 'The West End Route between London (Paddington) and Winchester and Southampton.' Three daily expresses ran via Newbury with 'Through Great Western carriages of the most improved description.' The DNS gained little from being part of a route six miles shorter between Didcot and Southampton than via Reading and Basingstoke until World War II when it became of tremendous importance, being doubled between Didcot and Newbury 18 April 1943 (more than a year ahead of D-Day). Improved track capacity was achieved by a variety of works south of Newbury in the same period. After the war it fell back into being a pleasantly rural line with intermediate stops like Churn, a platform seven miles south of Didcot. A timetable footnote stated that trains stopped to take up and set down on previous notice to ᵗhe stationmaster at Didcot, but warned that 'Evening trains call during daylight only.'

Passenger traffic continued to fall away and Newbury–Winchester services were withdrawn 7 March 1960, followed by Didcot–Newbury from 10 September 1962. Before then passengers with rail tickets were able to use a Sunday afternoon bus from Didcot, which returned from Newbury in early evening. Total closure followed in 1964 and the way was clear for improvements on the A34 main road between the Midlands and Southampton.

Besides its main and secondary lines, Newbury was once the terminus of trains on the twelve-mile LAMBOURN VALLEY branch. The northern half from Welford Park to Lambourn closed in 1960, but the remainder, serving a United States Air Force depot, survived until 5 November 1973, when the Western Region turned the occasion into a revenue-earning exercise, running four special trains for 3,000 enthusiasts and publishing a 'scrapbook.' There was never strong demand for a railway in the valley renowned for its racehorses and although Parliament approved a privately-promoted light railway in 1883, the first sod was not turned until 1888, and it was a further decade before trains ran from 2 April 1898. Seven years later the GWR took control and in a bid to meet road competition of the 1930s, tested a diesel railcar. It was so successful that one was built specially for the branch, fitted to meet the need for mixed trains by being fitted to pull a wagon or horsebox. The American base was served by a branch from the branch, for it ran three miles through the countryside from Welford Park.

A more important branch was that which the GWR built to the thriving town of MARLBOROUGH. As construction of the Berks & Hants Extension Railway went ahead, the GWR prepared to block any moves for a direct line between Southampton and the North by taking a controlling interest in the Marlborough Railway Company of 1861.

Opened broad gauge from Savernake on 14 April 1864, it was heavily graded, more than a fifth of its 5½ miles being at 1 in 58. The branch was converted in 1874, but its terminus remained a dead-end because the Swindon Marlborough & Andover Railway of 1881 approached on a broad curve to avoid the need to tunnel, and the only site for a through station lay a ¼-mile north. The branch was useful for local people, trains making Paddington connections at

Savernake. Grouping paved the way for economy and duplication between Marlborough and Savernake was ended on 6 March 1933 by a complicated scheme, adapting parts of both lines. The GWR station closed to passengers, but was retained for freight trains, using a shunt-back.

MIDLAND & SOUTH WESTERN JUNCTION RAILWAY

A slightly more explicit title for the M & SWJ might have been Midland & *London* & South Western Junction for it was an important link between those companies, despite opposition by the GWR, which fought for years to try to stop the building and later the development of through services on this line, which it regarded as a dangerous intruder in territory it thought traditionally its own.

But the independent company flourished, and in 1908 was advertising itself in assertive terms: 'This important Line, by its Through Service of Express Trains from Southampton in connection with the whole of the Midland Railway System, forms the Shortest, Quickest and most Direct Route between Paris, via Havre, French Coast, Channel Islands, Southampton, Portsmouth and the Midlands, North of England and Scotland.' Yet its heaviest traffic probably came from its role as a military line serving large army camps built on and around Salisbury Plain from late Victorian years.

The route was forged by the amalgamation of two local railways on 23 June 1884: the Swindon Marlborough & Andover of 1873, and the Swindon & Cheltenham Extension Railway of 1881. The SM & A reached Marlborough from Swindon (12½ miles) on 26 July 1881, joined the GWR at Swindon via Rushey Platt Junction on 6 February 1882, and on 1 May opened an isolated section from the Berks & Hants Extension at Grafton to the L & SWR at Red Post Junction (with running powers over the short distance to Andover). The Swindon & Cheltenham Extension was opened from Rushey Platt to Cirencester on 18 December 1883, the remaining 13¾ miles across the Cotswolds to the GWR Banbury–Cheltenham line at Andoversford Junction not being fully completed until 1 August 1891.

The route was finished and a through passenger service established over the ninety-seven miles between Cheltenham and Southampton, but the situation was difficult because of the GWR delaying trains. The M & SWJ therefore promoted (through a subsidiary company since it was then bankrupt) the 6¾-mile Marlborough & Grafton Railway, opened 26 June 1898. It bridged the West of England main line at Savernake, where the local company built a High Level station. The GWR had fiercely opposed the line, but support of the War Office proved overwhelming. Later its influence led the building of an east–south curve at Grafton so that troop trains could use the M & SWJ to reach Tidworth, headquarters of the Salisbury Plain Army District. The curve, double tracked, opened 6 September 1905.

Now it is necessary to go back some years to see how the M & SWJ was transformed in the 1890s from its original rather primitive state with little traffic potential from the thinly-populated countryside. The man responsible was Sam (later Sir Sam) Fay, who moved from the L & SWR as general manager in 1892. Apart from getting it out of bankruptcy, he developed through traffic, exercising running powers to Southampton for goods from 1892 and passengers two years later, establishing locomotive, carriage and wagon works at Cirencester for a motley collection of locomotives and rolling stock (coaches up to 51ft 6in), all painted in crimson lake livery; building the Marlborough & Grafton Railway, and establishing friendly relations with several companies.

Its major failure was in being unable to gain route independence north of Andoversford, and having to rely on the GWR between there and Cheltenham. Its first attempt to build north was when it found an ally in the Birmingham & North Warwickshire & Stratford-upon-Avon Railway, incorporated in 1894 to carve a path through Warwickshire independent of the GWR. The company lacked substance and had not started construction when in 1898 Parliament considered proposals for the Andoversford & Stratford-upon-Avon Railway, to be worked by the M & SWJ and connected to the Midland's Bristol–Birmingham main line. The plan was rejected when the GWR agreed to build between Cheltenham and Honeybourne to complete a through route to Stratford considered more useful.

The M & SWJ also considered a fourteen-mile link from Andoversford via Winchcombe to the Midland just south of Ashchurch, but abandoned the idea after talks with the Midland and GWR led to Paddington doubling Cheltenham (Lansdown)–Andoversford Junctions in 1900. It was conditional originally on similar work south of Cirencester, funded by a loan from the Midland, which in 1899 received running powers over the M & SWJ. The doubling never took place.

LATER YEARS : MIXED FORTUNES

At Grouping, the GWR obtained the running powers to Southampton that it had always wanted, but now they were of little value. The cross-country nature of the route made it susceptible to road competition and it fell into rural limbo, Andoversford–Cirencester being singled in 1928. World War II gave the line a tremendous boost, not least because of its army camps, but afterwards traffic fell dramatically and passenger services between Cheltenham and Southampton did not survive even until Beeching, being withdrawn 11 September 1961, and most of the line closed, apart from isolated sections retained for goods, including Savernake–Marlborough (closed 7 September 1964–the GWR High Level goods there having closed 19 May), and Swindon Town–Cirencester (Watermoor). The section from Swindon to Moredon Power Station survived until 1975 and in 1980 the Swindon & Cricklade Railway Society was reported to have won planning permission for a revived line between Moredon and Cricklade.

The historian T. B. Sands, who died in 1980, wrote a valediction on the line in the *Railway Magazine*, November 1961, noting that 'in its own small way, and for all too brief a period, [it] shone out as an example of railway enterprise at its best.'

CHAPTER XI

From Broad Gauge to HST

Tradition lives on. That was proved by at least one of the officials
looking after guests on the VIP special run a few days before the
introduction of High-speed trains between Paddington and Bristol
and South Wales on 4 October 1976. For prominent beside his BR
badges was that of the Great Western. HSTs quickly achieved
passenger popularity despite the service having been launched
during rising inflation and rail fares and they raised passenger
figures by a third in three years. Since them HSTs have spread to
other Western Region routes including the Berks & Hants, re-
signalled for high-speed running in the late 1970s having, retained
semaphore signals rather longer than comparable main lines in
other parts of Britain.

The HSTs are not popular with everyone: health officials at
Reading once complained that after they boarded a train to
inspect the kitchens, they had only three minutes before it reached
the limit of their jurisdiction at the borough boundary!

Bristol–Paddington HSTs (some starting from Weston-super-
Mare about thirty minutes earlier) and calling at Bath Spa, Swindon
or Reading, take just over ninety minutes. Passengers using Bristol
Parkway can reach Paddington almost fifteen minutes quicker than
from Temple Meads, and certainly quicker than on congested
motorways. Speed has proved the spur to Inter-City service
developments and having been privileged to make a 130mph-plus
trial run on the prototype HST between York and Darlington, I
know something of the efforts BR makes to meets motorway
competition.

The London–Bristol main line, although far shorter than the

Anglo-Scottish routes and lacking their operating difficulties like Shap, has been the setting of countless high-speed and occasional record-breaking runs.

More than a century earlier, notable runs came from Daniel Gooch's splendid locomotives (which contributed so much to broad gauge glory), when they were put to work to demonstrate its superiority. The 7ft 0in single *Ixion* reached 61mph between Didcot and London in June 1845, and three years later the new single *Great Britain* with 8ft 0in wheels, attained 78mph. It was an era in which the GWR was running by far the fastest trains in the world. But not for everybody. Third-class passengers were carried in open trains taking nine to ten *hours*, until the enforced introduction of Parliamentary services with covered accommodation from November 1844. The GWR compliance with the Act meant one 'Cheap Train' in each direction, while the night goods continued to carry third-class passengers.

Only gradually did comfort and speed improve. The Night Mail of January 1845 took 4 hours 20 minutes to Bristol and almost another 3 hours to Exeter. London–Bristol–West of England timings improved to meet both public demand and possible competition in the early 1860s when the L & SWR reached Exeter almost half-an-hour quicker. It was at a time when Parliament was debating its Bill for a competitive route to Bristol. Paddington–Bristol timings were down to two hours 15 minutes when the last broad gauge train ran on the day the gauge was abolished on 20 May 1892. The timings included a ten-minute enforced stop at Swindon for refreshments because of the condition of lease on the station refreshments rooms—a situation abolished in 1895. See *Railway Magazine* December 1980.

In 1895, the GWR Locomotive Superintendent, William Dean, wrote to John Pendleton, author of *Our Railways* outlining London–Swindon services. There were nine daily, running at an average speed of 53¼mph with loads of five to ten eight-wheel coaches.

For years some Bristol expresses carried a through portion for Avonmouth–West Indies banana boat passengers. The portion was detached at Temple Meads and usually with a coach added, worked to just short of Avonmouth Dock station, where a PBA locomotive took over for the short journey on dock lines to the customs shed.

Seafarers of earlier generations used the South Wales main line, as George Borrow recalled in the closing chapter of his classic *Wild Wales*, in which he relates travelling back home from Chepstow to Paddington on a night train in 1854. His most uproarious set of neighbours, a few carriages behind his own first-class one, were some 150 of Napier's tars returning from a Baltic expedition.

As the Bristol main line served places far beyond the City, so too did that between Paddington, Gloucester and South Wales, being constantly improved to meet growing Southern Ireland traffic. Yet it was four years after the opening of the Severn Tunnel in 1886 before most expresses were diverted via Bath, knocking up to thirty minutes off Gloucester timings with Cardiff four hours from London. Further speed-ups, this time immediately implemented, were achieved with the opening of the Badminton cut-off in 1903.

That brought welcome relief to Bristol's congested main lines, which carried Paddington–West expresses for another three years until the Exeter direct route opened, allowing the introduction of a service via the Berks & Hants which broadly has since changed little.

NORTH TO WEST: THE WELSH BORDER ROUTE

Bristol's long-distance traffic also came (and still does to a limited extent) from the North-to-West route, one which still delights, not only because of restored steam and the beauty of its countryside, but also with memories of its once infinite variety of through services. Take 1910 and the 9.00am express from Temple Meads, which called four minutes later at Stapleton Road, important and busy for decades. The express reached Shrewsbury at 11.49am, and portions reached Liverpool and Manchester (London Road) at 1.35pm. Connections to Leeds, York and Newcastle were offered to passengers who crossed 'the City of Manchester at their own expense.' The timetables did not stipulate whether they joined trains at Manchester Victoria or Exchange. Additionally, the Bristol express carried through coaches which, after shunting at Crewe, reached Glasgow (Central) and Edinburgh (Princes Street) at 6.00pm, with connections to Aberdeen (10.15pm).

Such long-distance expresses were supplemented in the Border country by Shrewsbury–Cardiff–Swansea trains. (Recommended further reading: Revd Keith Beck in *Railway World*, September 1980). Today, the route is an important secondary line, not least because it provides the only rail link (via Chester) between North and South Wales. The northern tip from Craven Arms carries extra traffic because of the reluctance of successive governments to close the Central Wales line, one of the oft-stated reasons being that it runs through six marginal Parliamentary constituencies. But main line locomotives were barred from the $78\frac{1}{2}$ miles–'crucial' miles was the term used by the NUR–between Craven Arms and Pantyffynnon from 1 January 1981. The move was made to delay the need for track renewal for a year or two.

BRISTOL–BIRMINGHAM: MIDLAND

The Midland could only offer West of England connections at Bristol for through coach services like that overnight from Glasgow (St Enoch) in 1910. It departed at 5.30pm, reached Birmingham (New Street) at 3.25am and was allowed only 2 hours 20 minutes to Temple Meads, including stops at Cheltenham (Lansdown), Gloucester and Mangotsfield. Bath (Queen Square) was reached at 6.47am. Day expresses ran shorter distances to Bristol, mainly from Bradford and Leeds. Timings varied little through the years. LMS expresses of 1938 took 2 hours 38 minutes including the same three stops between Bristol and Birmingham.

Major improvement did not come until Cardiff/Bristol/Birmingham–North services were completely revised on 10 September 1962, a timetable leaflet stating: 'As a result of more high powered Diesel Locomotives becoming available many improvements, both in service and journey times will be possible. For your convenience "The Cornishman" and trains between South Wales and Birmingham will serve New Street station instead of Snow Hill. In addition, "The Devonian" will run non-stop between Bristol and Birmingham.' One yardstick of the improvements was the timing of The Cornishman in 1 hour 56 minutes, a saving of 38 minutes.

Three years later the timetable was partly decimated by the withdrawal of local services (which Dr Beeching had recommended

only for modification) from 4 January 1965. Those between Bristol
and Gloucester were taking nearly 1 hour 30 minutes, those between
Gloucester and Birmingham (New Street) about 2 hours 15
minutes, much as they had done in Midland days.

BRISTOL–BIRMINGHAM: GWR

The GWR never introduced a local service between Bristol and
Birmingham after its route opened in 1908. It was developed by one
Bristol–Wolverhampton service on weekdays, increased to three the
following year. The Shakespeare Express in 1910 took 2 hours 21
minutes between Temple Meads and Snow Hill, with stops at
Stapleton Road, Cheltenham Spa (Malvern Road) and Stratford-
on-Avon—a creditable time for the $98\frac{3}{4}$ miles, ten more than the
Midland. It was the only one of the three trains to run beyond
Wolverhampton, terminating at Birkenhead. As Shrewsbury had a
a better service with North-to-West expresses, the GWR's catch-
ment area for its Bristol–Birmingham stem was confined to Birmingham and the West Midlands.

The same was true of Birmingham–Cardiff services, which came
into prominence when diesel railcars with buffets were introduced
on 9 July 1934, a supplement of 2s 6d ($12\frac{1}{2}$ new pence) on the
third-class fare giving a prestige touch for the first year. Businessmen responded, but by attracting capacity loadings, the railcars
showed their weakness of being unable to cope with heavy or
fluctuating demands and the railcars, which called at Cheltenham
(Malvern Road), Gloucester and Newport, had to be replaced
with steam expresses.

PADDINGTON–BIRMINGHAM

Far more important to the GWR than its Bristol–Birmingham
route was the London—Birmingham direct line established via
Aynho in 1910. In the Company's words: 'It gave the shortest
route between London and Birmingham, Wolverhampton, and
Shrewsbury and enabled considerable accelerations to be made in
the service to Chester, Birkenhead etc.' Commencing on July 1st
1910, a service of express trains between London, Birmingham and

Birkenhead by this route was inaugurated, the journey between London and Birmingham being covered in 2 hours, Shrewsbury 2 hours 59 minutes, Chester 4 hours 13 minutes and Birkenhead in 4 hours 37 minutes.

What the statement in the *Railway Year Book* of the period omitted was any reference to the Aynho route giving the GWR one to Birmingham 2.3 miles shorter than the LNWR's from Euston. It remained competitive until the West Coast electrification in 1967. That put an end to Paddington–Birkenhead (Woodside) expresses and left Shrewsbury without an Inter-City service, though after much protest—in Parliament and out—a through service to London, this time Euston, was restored. Expresses now take about 2¾ hours, some 20 minutes faster than in steam days. The Birmingham route is now Shrewsbury's only Inter-City one: those to Chester, Crewe and South Wales via Newport and the Central Wales being secondary rated.

LONDON TO GLOUCESTER, WORCESTER, HEREFORD

The three lovely cities where the Three Choirs Festival has been held each summer for more than two centuries are linked to Paddington by two lines that have lost status in recent years and been singled for part of their way across the Cotswolds. The Cheltenham and Great Western Union became famous after the introduction of Cheltenham Flyer on 9 July 1923, which a decade later gave Britain for a short time the 'World's fastest regular steam train.' However, the high-speed stretch was not across the Cotswolds, but between Swindon and Paddington, covered in 65 minutes at a start-to-stop average of 71.3mph. More recently, the Cheltenham Spa Express publicised the route, but the title was withdrawn when the train was decelerated. Today Cheltenham can be reached from Paddington in about 2 hours 20 minutes with stops at Swindon, Kemble, Stroud and Gloucester—about the level of the fastest timings before World War II.

When Paddington–Worcester expresses were accelerated to two hours with two stops in summer 1963 (and Kidderminster and Stourbridge portions withdrawn), they got back to 1914 schedules, though they have been decelerated since to give a better service to Cotswold stations that have survived, sustained by Oxford commuters and wealthy London-based executives. Estate agents give large houses and Cotswold estates rail appeal for buyers: 'Oxford-shire–near Burford . . . Charlbury station 5 miles (Paddington 80 minutes) . . .'

Some Worcester expresses continued via Malvern to Hereford, but while the 120 miles from Paddington to Worcester can be reached in about 2½ hours, the journey beyond dissolves into an all-stations marathon, the 15.00 from Paddington being allowed forty-eight minutes for the twenty-nine miles between Worcester and Hereford with five stops, a timing which helps to explain why Hereford passengers prefer travelling to Newport to catch HSTs.

CROSS-COUNTRY SERVICES

In the golden age of railways, the region was enmeshed by cross-country services developed by companies in fierce competition. The main flows were between the North of England, Scotland and the South West and South Wales. They still are–as a study of Table 51 in BR's annual passenger timetable and its own map quickly shows. Arguably the biggest change has been in the down-grading of the Shrewsbury & Hereford, otherwise the original main network is virtually intact, the casualties of economies having been the secondary lines.

Oxford was—and is—a fascinating centre of cross-country routes. Take summer 1896 when one could watch (or catch) a procession of trains with through carriages. They began with the 10.57am to Basingstoke, 11.12am Weymouth, 11.13am Liverpool (Central), 11.49am Aberystwyth, 12.22pm Southampton Docks, 1.00pm Bath, Bristol, Exeter, Kingswear, Taunton, Torquay (the listed order), and finally 2.17pm Basingstoke, Bournemouth (East and West), Eastleigh, (for Stokes Bay, Portsmouth and the Isle of Wight), Southampton West.

In later days, Oxford's chief popularity was among enthusiasts; it was an operators' nightmare, a bottleneck in which many through trains were caught. It became 'notorious for its inadequate facilities,' to quote one writer, R. Eckersley, who examined the problem in the *Railway Magazine* in February 1954.

Oxford is served by North-West–Gatwick–Brighton services,

which also give Reading a link with London's airports additional to the well-established Heathrow 'Railair' link by coach, which by 1973 was carrying more than 250,000 passengers annually. Reading station reception lounge had to be extended.

Bristol retains two long-distance cross-country services: Weymouth (eighty-seven miles) and Portsmouth Harbour (103 miles) on the Severn–Solent service, accelerated and intensified in 1979, when three more trains were introduced between Cardiff and Southampton. Bristol–Newport–Cardiff is a link maintained by trains taking about fifty minutes for the thirty-eight miles, slightly longer if calling at six intermediate stations.

Some of the many cross-country services that have gone were noted in the *Railway World* for October 1980 by E. C. B. Thornton, who one was delighted to see writing once more with splendid clarity about the railway complications of Gloucester and Cheltenham, after a lapse of seventeen years. He stated that for years Lansdown Junction, Cheltenham was the focal point from which trains or through coaches radiated to almost every part of Great Britain. They included those using the M & SWJ, the Somerset & Dorset, and Banbury & Cheltenham.

DEVELOPMENT OF SUBURBAN & COMMUTER SERVICES

For a city the size of Bristol, its local passenger network was remarkably small. The most intensive services of some forty weekday trains each way was developed by the Midland and GWR to Avonmouth from Bristol, some running to and from Bath via Mangotsfield; a few terminated at Stapleton Road. That was the broad pattern for about forty years as the area grew more and more to rely on buses, trams and cars. Such was the dominance they achieved that in 1962 the Western Region withdrew about 700 weekly services, heavily in the red. Subsequently the Severn Beach line was further pruned.

No such economies have taken place to the north and west of London, from which the Western Region operates its 'Commuteroute' out of Paddington with railheads at Windsor, Marlow, Henley, Reading, Newbury, Oxford, and from Marylebone to Princes Risborough. They form an extended version of 'Rural London' as the GWR defined the area in a 1909 *Guide to The*

Chalfont Country and The Thames Valley. It was subtitled 'Their Historic Landmarks and Residential Advantages.' It regarded the areas as ones which it 'has brought within easy reach of those who may need a brief period of rest or change, or be in quest of a permanent residence, away from the din and smoke of London, but within a convenient distance of the scene of their labours. The term (Rural London) applies equally well to Reading.'

The GWR tried the same residential hard-sell in South Birmingham's 'Beautiful Borderlands' (Vol 7), but never with the same success. Today, Birmingham's commuter routes include those to Stratford-upon-Avon and Leamington Spa, Worcester via Kidderminster, and less frequently via Bromsgrove which, like Redditch, is on the outer edge of the Cross-City service between Longbridge and Lichfield. In 1959–60, the Western Region campaigned hard to publicise its new diesel train services, more wide-ranging than today, for they included the Stourport loop via Bewdley. Perhaps with a sense of history, a pocket timetable gave pride of place to Birmingham–South Wales express diesel services of which 'Speed, comfort and cleanliness' were special features.

TRAINS—PIONEER AND FUNERAL

Motorail—a word that has now passed into the dictionary—was pioneered by three daily car-carrying trains through the Severn Tunnel between Pilning (High Level) and Severn Tunnel Junction. They ran from 1924 until 6 October 1966, nearly a month after the opening of the Severn road bridge, which also displaced the trains' earlier competitor, the Aust car ferry. Several motorail services run through the Thames & Severn Region, although two short-lived ones were withdrawn from 1978: Reading and Worcester to the West.

Paddington–Slough–Windsor has been the route of Royal funeral trains over the years and Sir Winston Churchill's funeral train to Handborough on 30 January 1965 ran from Waterloo via Ascot and Reading, hauled by a West Country Pacific bearing his name.

TRAIN - BUS COMPETITION

As bus competition developed, country branch lines were immediately vulnerable, but the GWR fought hard to keep and intensify
demand. In 1910 the GWR had two buses operating from Marlborough. One ran twice daily from *The Ailesbury Arms*, via the
station, to Calne, a 90-minute journey through Yatesbury. The
other was from Marlborough to Hungerford, via Ramsbury. The
buses ran only three times a week, and on Thursdays the first
service from Hungerford station would wait for the 11.3am (*sic*–
11.03am in actual fact) from Newbury, if passengers gave notice
before leaving there. In the same year, fourteen daily buses were
running on weekdays between Slough, Eton and Windsor, and
three between Wolverhampton and Bridgnorth, while in the Severn
Valley the GWR was reporting 'very keen competition' from
Midland Red buses. The GWR made its presence felt at Bromsgrove
with twice-daily buses to Stourbridge.

In the thinly-populated Cotswolds, buses from Stroud and
Painswick were extended to Cheltenham on market days; the GWR
also ran a Cheltenham–Winchcombe service until town and village
were connected by rail. Cheltenham was the base of an ambitious
GWR experiment in the 1930s, St James' station becoming the
terminal of 'Rail Motor Cars' on a Gloucester shuttle service, calling
at Churchdown, and also of Road Motors. They operated an
unusual through service to Paddington via Oxford, where bus/rail
transfer took place. Buses took two hours to cross the hills calling,
among other places, at Northleach and a stop in the village stated
to be (The Cotswold Press).

By summer 1935, the Hereford *ABC Railway Guide* freely acknowledged bus presence by prefacing train services with several
pages of rival bus services on every route out of the city. Nowhere
better was the growing superiority of local bus services demonstrated than in the Golden Valley. Because of lengthy connections,
railway journeys into the valley took about two hours—three times
as long as the direct buses.

The case for the continued existence of bus services and trains
was argued by the Railway Reinvigoration Society in a 1977
booklet *Can Bus Replace Train?* It examined the alternatives between
Hereford and Worcester. The findings were that the quickest bus

service, by Bromyard (long rail-less) took eighty minutes, while two other services took ninety minutes, twice as long as trains.

FREIGHT SERVICES

Ships came from all over the world with cargoes as varied as their ports of call: bananas by the million, fruits, hides, oil, petroleum (almost 100,000 tons in 1908), brandy, rum of the Indies, tobacco (very important), and timber. Avonmouth and Portishead (taken over by Bristol Corporation 1 September 1884) received them all. And away went a tremendous variety of exports. The first ship to sail from the Royal Edward Dock in 1908 had Bath bricks among its Australian cargo; the first ship in brought 15,000 cases of canned fruit from San Francisco (the first shipped direct to Bristol). It sailed the same evening loaded with tinplates. Foreign and coastwise tonnage handled at Avonmouth and Portishead passed the million mark in 1873 and doubled in the next thirty years. Trade build up began in the 1860s and the railways provided extra facilities as traded expanded.

In 1871—a year ahead of the withdrawal of the broad gauge in the Gloucester area—the GWR began a Bristol–Gloucester freight service, using running powers over the Midland route. It thrived and was subsequently switched via the Severn Tunnel.

At Bristol, the main lines of both the big companies fed into sorting yards and dozens of sidings. About a quarter of these belonged to the Midland. They served a host of installations, for besides the main docks there were ship repair yards and the private wharves of engineering works, oil and petrol depots, gas works, several tobacco factories. Bristol Corporation actually maintained its own tobacco siding, and also another for cattle. At Pylle Hill GWR goods yard, the firm of Pooley & Son had a depot for serving railway-owned weighing machines at stations from a wide area.

Highest priority went to the Avonmouth banana traffic, being worked to London and elsewhere by special trains. As the GWR *Appendix to Service Time Tables* stated in 1932: 'Everybody concerned must give special attention to the running of the Special Trains, and with the exception of Passenger Trains, they must have precedence over all others . . . Signalmen must advise the Specials forward on the Box to Box telephone.'

The next instruction dealt with livestock. 'Live Stock Traffic must not be accepted from the L.M. & S Company at Bristol unless there is a forward service within about three hours of the transfer being made.'

Some of the heaviest freight flows through the region were of South Wales coal destined to all parts of Britain. Much was routed via Bristol, rather than Gloucester, once the Severn Tunnel opened, while it went north over a variety of routes including the Heads of the Valleys line, Abergavenny, and the Newport–Shrewsbury route, which received further traffic off the Midland's Swansea route at Hereford.

The Midland's Bristol–Birmingham route carried a tremendous volume of freight destined to and from a number of routes, including the Somerset & Dorset, Swindon–Gloucester–Cheltenham. The Birmingham & Gloucester loop and the Banbury–Cheltenham line were used to keep freight clear of the heart of the West Midlands, especially important in wartime.

Beside being a major junction for freight traffic, Gloucester generated a fair volume of its own from the docks, used by coasters of up to about 1,000 tons, which carried coal, petroleum, and timber. Lack of freight lines to avoid Gloucester Central station led to congestion, so locomotives of goods trains from South Wales bound for London often had to take water twice to avoid losing paths between passenger services. Paths had also to be found through Central for transfer trips between local yards, including the 'T' Sidings, the Old Yard and the GWR Docks branch, where rails across a swing bridge formed the shunting neck into Llanthony Yard.

On Nationalisation, the mainly GWR goods traffic at Reading was merged with that of the Southern, Central Goods station in Vastern Road (near to Reading General) becoming the railhead for an area of about 1,000 square miles. In the early 1960s it handled about 500,000 tons of coal a year, 135,000 tons of general goods and up to 1,000 trucks of livestock. A staff of 120 dispatched about 50,000 tons of general goods, including coke, tar, machinery, farm produce and beer for export. Biscuits and cakes, a traditional Reading industry, were sent from Huntley & Palmer's private sidings. The Central (Coley) Goods Depot continued to deal with a variety of traffic, including coal, fertilisers, oil and timber.

High Wycombe's freight traffic included furniture. Towns like Kidderminster, which once advertised itself with 'carpet weaving—staple industry' frequently used good communications as a sales point in campaigns to attract new industry. It said, for instance, that there was speedy and efficient transport by the GWR and LMS 'at moderate freight charges.' The partial re-opening of the Thornbury branch for quarry traffic has been noted earlier (page 56).

CHANGING TRAFFIC PATTERNS

Branch closures led to the concentration of new traffic on larger centres. In 1974, Shrewsbury began handling farm balers being sent to France from a new factory at Market Drayton, rail-less since the closure of the Stoke-on-Trent branch and the Crewe–Wellington line. A year earlier, BR won its first contract for the short haul (seventy miles) of car components in bulk from Swindon to Longbridge. Cars were already being railed from Abingdon and Cowley to several destinations, including ports.

But attempts to develop car traffic from Cowley through a privately-developed container terminal and distribution centre at Didcot were unsuccessful because of inter-union strife. Southampton dockers claimed they should handle the container traffic at this 'inland port.' And there was pressure for British Leyland to distribute cars by road because it was claimed one train would take away the jobs of twenty lorry drivers. As the Transport Salaried Staffs' Association commented in its *Journal:* 'That is what the whole issue has been about–people's jobs.' It closed in 1977, but was re-opened without rail car-handling facilities by new owners in June 1981.

Meanwhile the railways in the Didcot area carried coal from Nottinghamshire to the large power station using one of the longest merry-go-round train circuits in Britain, and in handling London's waste being dumped at Appleford to reclaim farm land after gravel pits became worked out.

ACCIDENTS

One of the earliest accidents has turned out to be among the best-remembered: the boiler explosion of a Birmingham &

Gloucester locomotive in Bromsgrove station on 10 November 1840, which killed the driver and fireman. Their inscribed graves in the local churchyard have been restored in recent years—and vandalised.

A far more serious accident happened at Shipton-on-Cherwell, near Woodstock on Christmas Eve 1874 when thirty-four people were killed and sixty-nine injured after the front coach of a Paddington–Birkenhead express fractured a tyre.

Another crash which attracted national attention occurred at Shrewsbury on 15 October 1907 when the Crewe–West of England night mail ran through danger signals at full speed and round the tight curve at the station approach. It piled up inside the station, killing eighteen people, including eleven passengers. There was no certain explanation, but there was strong public reaction because it was the third accident of its kind in just over a year, following Salisbury and Grantham.

There were two serious crashes on the Bristol–Birmingham line in the late 1920s. Fire, caused by gas escaping from coaches, wrought havoc at Charfield on 13 October 1928. Fifteen people perished when a Leeds–Bristol mail over-ran danger signals on a misty morning and crashed into a GWR goods being shunted out of its path. Moments later it would have been in the loop, clear of the main line, but the express travelling at about 60mph hit the goods train and cannonaded into another bound for Gloucester on the up line. Wreckage, piled high in a deep cutting, blazed for twelve hours. Two of the dead were young children whose bodies were charred beyond recognition. No-one ever came forward to identify them.

An almost carbon copy of the Charfield accident occurred at Ashchurch on 8 January 1929 when four people were killed after a Bristol–Leeds express over-ran signals. It led Inspecting Officers to urge adoption by the LMS of the GWR automatic train control system, but it was only ever installed between Bristol and Abbot's Wood Junction, near Worcester. Some Joint lines, including those near Shrewsbury, had it, but only for use by GWR trains.

BR finally abandoned the GWR system amid controversy from Western men following the Region's first serious post-war crash at Milton, near Didcot, on 20 November 1955 when a Sunday excursion from Treherbert to Paddington was derailed while taking a crossover at high speed. Eleven people were killed, more than 160 injured. The accident caused disquiet because it happened on a main line where ATC had been used for years. Progressively, the ATC was replaced by BR's standard automatic warning system. The GWR had pioneered ATC on the Henley branch 1905–08. The last piece was removed at Stratford-upon-Avon in 1979.

PRIVATE LOCOMOTIVE BUILDERS

Flourishing railway supply industries were built up in several places. At Bristol, industrial locomotives far more varied in design than those used on the main lines, were produced by private builders for home and abroad.

The Atlas Engine Works at St George was established by Fox Walker & Company in 1864 and taken-over by Peckett & Company sixteen years later. The third and short-lived control from a loco-motive point of view came in 1961, when the firm passed to the Reed Crane & Hoist Company of Brighouse, Yorkshire. That was three years after the last steam locomotive had been produced, a 3ft 0in gauge tank for a Mozambique sugar estate. Peckett's policy was against diesel building for many years and only five were produced, all between 1956–59. The first, an 0–4–0, worked at the Portishead works of Albright & Wilson for a short time. The Peckett works was connected to the Gloucester line at Kingswood Junction by a longish branch, which also served East Bristol pits. It was not entirely satisfactory, for the overbridges were too low for the largest Peckett locomotives and they left the works partly dismantled. The branch closed in 1958. Another Bristol builder of industrial locomotives was the Avonside Engine Company.

Bristol had several firms associated with carriage and wagon building and hire. The Bristol & South Wales Railway Wagon Company of 1860 was purely a wagon owning company, while the Bristol Wagon & Carriage Works Ltd., had its headquarters in Lawrence Hill for many years.

Also dating from 1860, was the Gloucester Railway Carriage & Wagon Company, connected to the Midland's Tuffley Loop for many years. The Midland Railway Carriage & Wagon Company was at Birmingham and Abbey Works, Shrewsbury—a town where Rolls Royce build diesel locomotives in the former Sentinel Works.

Based in Bristol, but with its only works in Cardiff was the Western Wagon & Property Company, again established in 1860.

Thousands of signals have been manufactured at Chippenham, home of Westinghouse Company, and Worcester, where McKenzie & Holland—sole licensees for the Westinghouse Electro–Pneumatic system of interlocking—set up their head office and works. Dutton's was another well-known Worcester signalling firm, and the City was a locomotive building centre in the 1860s. The Worcester Engine Company supplied the North Staffordshire, Great Eastern and Bristol & Exeter Railways, as well as to a number of overseas companies. They included the Great Russian Railway.

MEMORIES OF STEAM

HSTs and steam locomotives rub shoulders at Didcot, where the Great Western Society is based. And a further link between the two, deliberately fostered is on BR's 'Journey Shrinker' leaflets, for Swindon is symbolised by one of the Dean 0–6–0 goods locomotives, which haunted many branches in deep England for years. Swindon is described as an important railway town with Georgian inns and a modern shopping centre.

Remaining Swindon locomotives, notably No 6000 *King George V*, have been active for several years on the Newport–Shrewsbury route, which at ninety-four miles, was the longest of the five re-opened to steam in 1972. The second longest was Birmingham (Moor Street)–Didcot, seventy-seven miles. The following year the two routes between Birmingham and Stratford-upon-Avon were re-opened, together with the eighty-five miles between Oxford–Worcester–Hereford, but it was withdrawn in 1975 because of operating problems over the single line.

Cheltenham was the birthplace in 1928 of The Railway Correspondence & Travel Society, which has since imbued many a young spotter with a love, understanding and respect for railways.

A few miles north of Cheltenham, the Dowty Railway Preservation Society was established at Ashchurch in sidings outside the factory in 1962, and locomotives and stock can be glimpsed from the Bristol–Birmingham main line.

Steam shed sites survive. After Bath Road locomotive depot (beside Temple Meads) was last used regularly by steam in 1960, it became a main line diesel depot. During alterations, well-preserved broad gauge lines dating back to 1892 were unearthed, having been buried since then. St Philip's Marsh shed on the Direct Line was demolished after closure in 1964 and an HST servicing depot built on the site and that of the goods depot, closed 1973. Part of the land was used for factory building and a four-road depot to service about 100 dmu cars was built near Marsh Junction. A new civil engineering depot for the area was established at Ashton Gate, replacing ageing ones at Pylle Hill, St Philip's Marsh and Bath.

The old GWR main lines remain intact apart from track and station economies. A touch of steam and the return of lower quadrant signals, and the scene could be that of many years ago.

APPENDIX

RAILWAYS OF BRISTOL AND AVONMOUTH

This list, which can be used in conjunction with the maps on pages 24 and 40 gives the main dates of railways in the Bristol and Avonmouth areas.

CHAPTER II FROM LONDON TO BRISTOL

Dates are restricted to the main line opening and Bristol area changes.

The Bristol & Exeter (Vol 1) was the first railway to open from Bristol (to Bridgwater on 14 June 1841) but not the first to be authorised. It was Incorporated on 19 May 1836, while the London & Bristol had been authorised 31 August 1835. It opened in nine comparatively short sections, of which the first was the longest.

4 June 1838   Paddington–Maidenhead. 22½ miles.

1 July 1839   Maidenhead 'Taplow'–Twyford. 8¼ miles.

30 March 1840   Twyford–Reading. 5 miles.

1 June 1840   Reading–Steventon. 20½ miles.

20 July 1840   Steventon–Faringdon Road (Uffington). 7¼ miles.

31 August 1840   Bristol–Bath. 11½ miles.

17 December 1840   Faringdon Road–Hay Lane. 16½ miles.

31 May 1841   Hay Lane–Chippenham. 13½ miles.

30 June 1841   Chippenham–Bath. 13 miles.

CENTRAL BRISTOL: LINE, STATION, SHED CHANGES

January 1852   Bristol & Exeter: Bath Road Locomotive Works opened.

19 June 1865   Bristol Joint Station Act: Reconstruction of Temple Meads.

6 July 1867   Bristol Temple Meads: New platforms opened.

1 January 1878   Bristol Temple Meads: Reconstruction completed.

10 April 1892   Bristol Relief Line opened, 1 mile. Also Marsh Junction (new connection with Frome branch).

9 July 1910   St Philip's Marsh locomotive depot opened.

6 September 1965   Temple Meads: Old train shed closed. Part converted to car park.

7 August 1967   Bristol East Yard (Up & Down yards) closed.

CHAPTER III   BRISTOL & AVONMOUTH

This section details lines between Bristol and Avonmouth and to docks at both these places, all part of the Port of Bristol.

BRISTOL HARBOUR RAILWAY

28 June 1866   Bristol Harbour Railway incorporated: Temple Meads–Wapping Wharf. ¾-mile.

11 March 1872   BHR opened.

6 August 1897   BHR: new lines authorised.

4 October 1906   BHER opened: Wapping Wharf–Ashton Junction opened, 1 mile. Also: Canon's Marsh Goods depot.

6 January 1964   Temple Meads–Wapping Wharf closed.

14 June 1965   Canon's Marsh Goods Depot and branch closed.

BRISTOL PORT RAILWAY & PIER

17 July 1862   Bristol Port Railway & Pier incorporated: Bristol Hotwells–Avonmouth. 5¾ miles.

6 March 1865   BPR & P opened.

15 August 1867   BPR & P (Clifton Extension) authorised: Sneyd Park–Ashley Hill. 3¼ miles.

25 May 1871   BPR & P. Transferred to GWR & Midland Railway jointly.

1 October 1874   CER: Opened Ashley Hill Junction—Clifton Down. 1½ miles.

22 February 1877   BPR & P open Avonmouth Dock.

24 February 1877   CER: Clifton Down–Sneyd Park opened, goods.

1 September 1885   Bristol (Temple Meads)–Avonmouth passenger service introduced by GWR. Avonmouth Dock station (GWR/Midland Joint) opened.

1 October 1886   MR: Bristol (St Philip's)–Avonmouth passenger services abandoned.

25 July 1890   BPR & P purchased by GWR and MR (from 1 September).

18 May 1903   Avonmouth: Terminus closed and passenger service cut back to Dock station.

9 July 1908   Royal Edward Dock opened, and new GWR/MR station.

16 August 1920   Bristol Corporation take over Hotwells–Sneyd Park Junction. 1¾ miles.

19 September 1921   Hotwells station closed. Line cut back to Hotwells Halt. ¼-mile through Gorge abandoned.

3 July 1922   Hotwells Halt–Sneyd Park Junction closed completely.

15 March 1948   London Midland Region (ex-Midland Railway) lines at Avonmouth transferred to Western Region.

10 July 1967   Avonmouth: Dock Junction–St. Andrew's Junction goods lines closed. 1 mile. Also Avonmouth Old Yard.

MIDLAND: KINGSWOOD–ASHLEY HILL JUNCTIONS

15 August 1867   Bristol Port Railway & Pier (Clifton Extension) Act: Sneyd Park–Ashley Hill. Connecting spurs to GWR and Midland.

1 October 1874   MR: Kingswood Junction–Ashley Hill Junction opened. 1¾ miles.

31 March 1941   LMS: Fish Ponds–Clifton Down closed to passengers.

14 June 1965   BR: Kingswood Junction–Ashley Hill Junction closed completely.

AVONMOUTH–PILNING

4 August 1890   GWR: Avonmouth–Pilning authorised. 7¾ miles, including 1¾ miles of New Passage line.

28 June 1892   Deviation: Gloucester Road–Holesmouth Junction authorised (for Royal Edward Dock construction).

5 February 1900   Avonmouth–Pilning Junction opened: Goods.

22 November 1903   Deviation completed.

23 June 1928   Avonmouth–Pilning opened to passengers. Station at Severn Beach.

9 July 1928   Halts opened: New Passage, Cross Hands, Pilning (Low Level).

27 October 1963   Branch: Pump House Siding–Sea Wall Pumping Station closed completely, One-third mile.

23 November 1964   Severn Beach–Pilning closed to passengers.
2¼ miles.

25 July 1968   Severn Beach–Pilning closed completely.

GWR: AVONMOUTH–STOKE GIFFORD

15 August 1904   Avonmouth (Holesmouth Junction)–Filton
Junction–Stoke Gifford authorised. 6¾ miles.

9 May 1910   Route opened. Henbury station and halts: Hallen,
Charlton and North Filton. Also Filton West curve. ¾-mile.

13 May 1917   Route doubling completed.

12 July 1926   North Filton platform opened. Replaced Filton
Halt, closed 1915.

23 November 1964   Bristol (TM)–Avonmouth via Filton Junction
passenger service withdrawn. Workmen's trains continued
between Bristol (TM) and North Filton until 5 September 1966.

22 May 1966 Avonmouth (Hallen Marsh Junction)–Filton West
Junction singled.

22 February 1971   Patchway–Filton West Junctions: new curve
opened.

AVONMOUTH LIGHT RAILWAY

12 December 1893   Avonmouth Light Railway incorporated:
Bristol Port Railway & Pier–Avonmouth to Filton line (later
Holesmouth Junction), 2 miles.

1 December 1903   Scheme revived, and again 26 February 1912.

1908   Construction begun.

1918   First ¾-mile completed. Remainder abandoned.

29 March 1927   ALR purchased jointly by LMS and GWR.

c. 1935   ALR closed.

CLIFTON ROCKS RAILWAY

1890   Clifton Rocks Railway promoted.

7 March 1891   Construction begun.

11 March 1893   Railway opened.

29 November 1912   CRR acquired by Bristol Tramways &
Carriage Company.

1 October 1934   CRR closed.

CHAPTER IV   FROM BRISTOL TO BIRMINGHAM

Bristol area tramroads later formed the route of the Midland Railway main line to Birmingham, dealt with here only to Gloucester.

19 June 1828   Bristol & Gloucestershire Railway incorporated. Tramway: Bristol (Cuckold's Pill, later Avon Street Wharf)–Coalpit Heath and collieries at Shortwood, Parkfield and Mangotsfield, 10 miles.

Avon & Gloucestershire Railway incorporated. Tramroad Avon Wharf, Bitton–Mangotsfield. 4½ miles.

5 July 1831   A & GR connected to B & GR Mangotsfield.

July 1832   B & GR: Northern section Mangotsfield–Coalpit Heath opened.

6 August 1835   B & GR opened throughout. Horse tramroad.

1 July 1839   B & GR: Conversion authorised to locomotive railway Bristol–Gloucester. GWR running powers Bristol–Standish Junction (with Cheltenham & Great Western Union). Name shortened to Bristol & Gloucester Railway.

13 April 1843   B & GR lease to GWR agreed.

1 July 1843   C & GWU absorbed by GWR.

8 July 1844   B & GR opened. 33 miles. Bristol and Birmingham now rail-linked. Goods: 2 September.

1 January 1845   B & GR and Birmingham & Gloucester Railways: Traffic 'worked as one' pending amalgamation with GWR.

7 May 1845   Midland Railway takes possession of Bristol & Gloucester and Birmingham & Gloucester Railways.

1 July 1845   Bristol & Gloucester leased to MR. Absorbed 3 August 1846.

1 July 1851   Kennet & Avon Canal Company (including A & GR) absorbed by GWR. (Railway abandoned by GWR Act 5 July 1865).

22 May 1854   Bristol (Temple Meads)–Gloucester. Start of standard gauge passenger service.

1 April 1866   Stapleton station opened. (Fish Ponds from 1 July 1867).

4 August 1869   Mangotsfield: new station opened for Bath branch.

2 May 1870   Bristol (St Philip's) station opened.

26 May 1872   Bristol–Gloucester. Broad gauge rails removed.

1881   Avon & Gloucester Railway: California Colliery–Avon Wharf, Keynsham re-opened. (All traffic ceased 1905).

1 November 1888   Staple Hill station opened.

9 March 1908   GWR: Westerleigh West–Yate: Junction with Midland Railway. 1¼ miles. Westerleigh East Loop. ½-mile.

10 July 1927   Westerleigh North–East curve closed. (Re-opened 16 August 1942 to 4 January 1950).

31 March 1941   Bristol: Kingswood–Ashley Hill Junctions closed to passengers. 1½ miles.

21 September 1953   Bristol: St. Philip's–Lawrence Hill Junction closed to passengers: services diverted to Temple Meads.

10 September 1962   Mangotsfield :South–North Junctions closed to passengers.

21 August 1964   Brookthorpe marshalling yard plan abandoned by Western Region.

8 November 1964   Bristol (St Philip's)–Lawrence Hill Junction: branch converted to siding. (New connection Lawrence Hill from 2 February 1970).

4 January 1965   Bristol (Temple Meads)–Gloucester (Eastgate) local passenger services withdrawn and stations closed: Yate, Wickwar, Charfield, Berkeley Road, Coaley Junction, Stonehouse (Bristol Road), Haresfield. (Frocester closed 11 December 1961).

22 February 1965   Westerleigh marshalling yard closed.

7 March 1966   Bristol (Temple Meads)–Mangotsfield–Bath (Green Park) passenger service withdrawn. Also: Somerset & Dorset Railway closed to passengers and virtually completely.

29 January 1967   Mangotsfield South–West Spur closed.

29 December 1969   Bristol (Temple Meads)–Yate closed to passengers. Lawrence Hill–Mangotsfield completely. Line singled north to Mangotsfield.

18 January 1970   Yate Fly-over closed and Gloucester line realigned.

28 May 1971   Mangotsfield–Bath: branch closed completely.

CHAPTER V   MAIN LINES TO SOUTH WALES

Section deals with two lines and the Severn Tunnel section, all integral parts of the Greater Bristol network.

BRISTOL & SOUTH WALES UNION

27 July 1857   Bristol & South Wales Union Railway incorporated: GWR ½-mile east of Temple Meads–New Passage Pier. 11½ miles. Also Portskewett Pier–South Wales main line, ½-mile. Broad gauge.

8 September 1863   B & SWU: Portskewett–Pier opened.

1 January 1868   B & SWU opened: Portskewett–Pier.

1 August 1868   B & SWU amalgamated with GWR.

11 May 1872   Portskewett–Pier converted broad to standard gauge. Remainder: 9 August 1873.

10 August 1885   Patchway: new station opened.

21 February 1886   Bristol: Dr. Day's Bridge Junction opened.

29 May 1886   Bristol Avoiding Line: North Somerset Junction–Dr Day's Bridge Junction opened, ¼-mile.

1 September 1886   Narroways Hill Junction–Patchway doubled.

1 December 1886   Portskewett Pier closed. Pilning Junction–New Passage Pier closed completely. (Partly re-opened 1900). Pilning (High Level) station opened.

27 May 1887   Patchway–Pilning doubled.

8 January 1888   Narroways Hill Junction closed and Avonmouth lines extended to Stapleton Road.

20 September 1891   North Somerset Junction–Stapleton Road quadrupled.

1 July 1903   Filton Junction: new station opened. 2¾ miles.

30 April 1933   Stapleton Road–Filton Junction quadrupled. 2¾ miles.

26 November 1964   Portskewett station closed.

6 May 1974   Bristol (Lawrence Hill) and Bristol (Stapleton Road) lose prefix "Bristol"

SEVERN TUNNEL

5 July 1865   South Wales & Great Western Direct Railway authorised: Wootton Bassett–Chepstow. 41 miles. Abandoned 1870.

27 June 1872   GWR: Pilning Junction–Severn Tunnel Junction authorised.

18 March 1873   Construction begun: Sudbrook (Monmouth). 7½ miles.

1 September 1886   Tunnel and line opened for goods; passengers 1 December 1886.

6 November 1961   Cornish beam pumping engine replaced by electric pumps.

SOUTH WALES & BRISTOL DIRECT RAILWAY

7 August 1896   South Wales & Bristol Direct Railway authorised: Wootton Bassett–Patchway & Filton: Westerleigh–spurs to Midland Railway at Yate. 33¼ miles. Also Berkeley Loop: Berkeley South–Berkeley Loop Junctions. 1¼ miles.

1 January 1903   Wootton Bassett–Badminton opened goods. 17 miles.

1 May 1903   Badminton–Patchway open to goods. 12½ miles. Also: Westerleigh West & East Junctions opened to goods, and Berkeley Loop. (Passengers 2 November).

1 July 1903   Wootton Bassett–Patchway opened to passengers.

3 April 1961   Wootton Bassett–Patchway. Intermediate stations closed except Badminton (closed 3 June 1968).

4 October 1971   Stoke Gifford marshalling yards closed.

1 May 1972   Bristol Parkway station opened.

Acknowledgments and Bibliography

My task of sorting out a complicated Region with many ragged ends has been made pleasant and easy through the willing help of many people. I am especially grateful to John Norris, who has written extensively on the canals and railways of the Region, for checking the manuscript and putting me back on the rails several times. If there are further derailments, the fault is mine alone. I am grateful to a colleague, Gerry Clarke, for photographic work and advice.

Other equally appreciated help has come from C. A. Appleton, G. Biddle, R. E. Barby, D. S. M. Barrie, B. Baylis, D. Bone, P. H. Edwards, C. Farmer, H. Forster, Dr J. R. Hollick, J. Lovelock, M. T. Hale, J. Mair, R. W. Miller (my co-author of other railway histories), G. Moss, M. P. N. Reading, L. Smith, N. Sprinks, E. S. Tonks and J. Williams.

Besides contemporary documents, minutes, timetables, travel guides, newspapers (including several centenary supplements), *The History of the Great Western Railway*, Volumes I & II by E. T. MacDermot and C. R. Clinker, has been my most thumbed source book. Clinker's *Register of Closed Passenger Stations & Goods Depots in England Scotland & Wales 1830–1977* is another praiseworthy work, as are several publications of the Branch Line Society, including its fortnightly Newsletter, *A Guide to Closed Railways in Britain 1948–75* by N. J. Hill and A. O. McDougall, *Branch Line Index* by G. C. Lewthwaite, and *Passenger Services over Unusual Lines*, by R. Hamilton and B. W. Rayner.

Journals of the Railway & Canal Historical Society, the Railway Correspondence & Travel Society and the Stephenson Locomotive

Society reflect the changing scene, together with several books, notably G. Freeman Allen's *The Western Since 1948*, *The West Midlands: A Regional Study;* guides to revived steam railways and centres; *Passengers No More*, by G. Daniels and L. Dench, *Closed Passenger Lines of Great Britain 1827–1947* (M. D. Greville & J. Spence), *Rail Atlas of Britain* (S. K. Baker).

Specialist aspects are covered by an increasing number of books: *An Historical Survey of Selected Great Western Stations* (R. H. Clarke), *Track Layout Diagrams* (R. A. Cooke), *LMS Engine Sheds* vol. 1 (C. Hawkins and G. Reeve), *An Historical Survey of Great Western Engine Sheds* (two volumes) (E. Lyons and E. Mountford), *A Pictorial Record of Great Western Architecture* (A. Vaughan); *Saving Railway Architecture* by SAVE.

Among railway histories consulted apart from those mentioned in the text, were *Britain's Joint Lines* (H. C. Casserley), also his evocative *History of The Lickey Incline; Great Central* (Vols I–III) by George Dow; *The Oxford Worcester & Wolverhampton Railway* (S. C. Jenkins & H. I. Quayle); *The Railway Mania & Its Aftermath* (H. G. Lewin); *The Railway in England & Wales 1830–1914* (Vol 1) by Jack Simmons.

Detailed histories of secondary, branch and minor railways and services consulted included those of *The Gloucester & Cheltenham Railway* (D. E. Bick), *GWR Branch Lines 1955–65* (C. J. Gammell), *The Watlington Branch* (J. Holden), *The Woodstock Branch* (R. Lingard) —also his *Princes Risborough Thame & Oxford Railway; The Bristol & Gloucester Railway, The Bristol Port & Pier Railway; The Midland & South Western Junction Railway* (C. Maggs), *Great Western London Suburban Services* (T. B. Peacock), The *Cleobury Mortimer & Ditton Priors Light Railway. The Lambourn Valley Railway* (M. R. C. Price), *The Banbury & Cheltenham Railway* (J. H. Russell); *The Didcot Newbury & Southampton Railway* (T. B. Sands); *The History of Gloucester Docks and its associated canals and railways* (M. Stimpson), *The Cleobury Mortimer & Ditton Priors Light Railway* (W. Smith & K. Beddoes).

Several Forgotten Railway books were helpful: *Chilterns & Cotswolds* (R. Davies & M. D. Grant) and *South Wales* (J. Page); and useful information concerning the growth of lines is to be found in *The Oxfordshire Landscape* (F. Emery) and *The Gloucestershire Landscape* (H. P. R. Finberg). Several canal books contain useful railway information including *The Hereford & Gloucester Canal* (D. E. Bick); *The Canals of South & South East England* (Charles Hadfield).

The short extract from *Brensham Village* by John Moore is reprinted by permission of A. D. Peters & Co. Ltd.

Index

Figures in **bold** denote illustrations